The Planets

Robin Kerrod

Silverdale Books

Published by SILVERDALE BOOKS
An imprint of Bookmart Ltd
Registered number 2372865
Trading as Bookmart Ltd
Blaby Road
Wigston
Leicester LE18 4SE

© 2007 Graham Beehag Books

Graham Beehag Books
Christchurch
Dorset

This edition printed 2007

ISBN 978-1-845094-35-5

The material in this book is revised and
previously appeared in the series Planet Library

Printed in Singapore

1 3 5 7 9 10 8 6 4 2

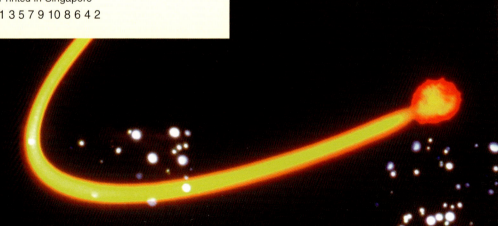

The
Planets

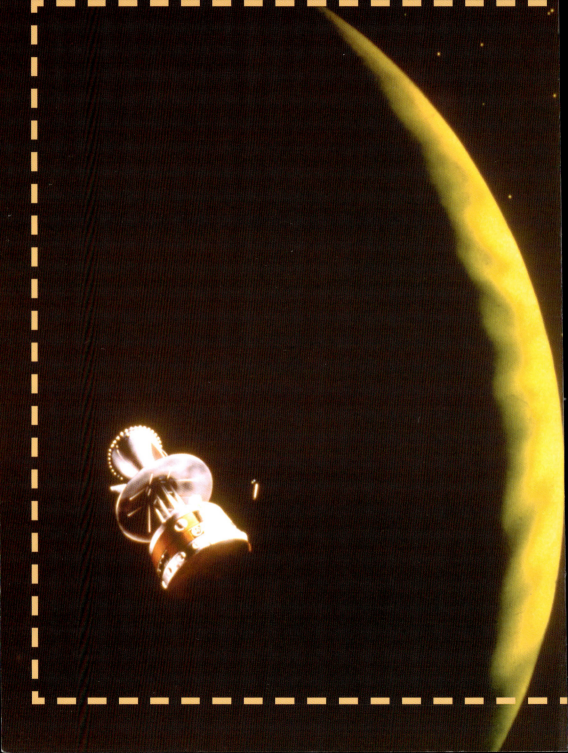

Contents

B

WARNING!

Never look directly at the Sun. In particular, never look at the Sun through binoculars or a telescope. If you do, you will damage your eyes and may go blind.

A streamer of glowing gas thousands of miles long shoots high above the Sun's surface. In time, the gas will cool and fall back into the Sun.

The Sun

To us on Earth, the Sun is by far the most important body in the heavens. It sends us light to see by and heat to keep us warm. Without the Sun's light and heat, plants could not grow. Then there would be no food for other living things. Earth would be a dark, cold, and dead world.

The Sun and Earth travel together through space. Earth is one of many bodies that circle around the Sun. Together, these bodies form the Sun's family, the solar system. The word solar comes from the Latin word for the Sun, *Sol*.

The Sun is the only body in the solar system that gives off its own light. Planets and moons only reflect light from the Sun. The Sun is a star. It looks bigger and brighter than the other stars in the night sky because it is much closer.

Like the other stars, the Sun is a huge ball of extremely hot glowing gas. Its glaring surface is stormy and always changing. Great fiery fountains suddenly spring up. Dark spots come and go. Streams of particles flow out into space like a wind, causing spectacular effects when they reach Earth.

Astronomers think that the Sun is 4.6 billion years old. They believe it will probably keep on shining as it does for another 5 billion years. Then it will start to die. In time it will fade and shrink to a dark ball of cinders not much bigger than Earth.

Symbol for the Sun

Our Star, the Sun

The Sun is quite unlike any other body in the solar system because it is a star. It makes its own energy and pours this energy into space as light and heat. But it is a very ordinary kind of star.

The Sun God

Ancient peoples knew how important the Sun was to them, and they worshiped it as a god. In ancient Egypt, for example, Re was worshiped as the Sun god and the creator of all things.

Like all stars, the Sun is a great ball of very hot gases. The main gas is called hydrogen. We know that the Sun is very hot because we can feel its heat a long way away, on Earth. Earth lies about 93 million miles (nearly 150 million km) away from the Sun.

In our skies, the Sun looks much bigger and very much brighter than the other stars. But as stars go, the Sun is not very big or very bright. It only seems bigger and brighter to us because it is much nearer. The nearest stars are not just millions of miles away, they are millions of millions of miles away. Imagine that our Sun was the size of a grain of sand and located in New York City. On the same scale, the nearest star would be another grain of sand located in San Francisco.

THE YELLOW DWARF

The Sun measures about 865,000 miles (1,400,000 km)

The Sun sinks below the western horizon at sunset. The sky turns red because the dusty atmosphere reflects red light from the Sun better than light of other colours.

supergiant star

Sun

Jupiter

Earth

in diameter. That is more than a hundred times bigger than Earth. But for a star, the Sun is actually small. Astronomers classify the Sun as a dwarf star. They call it a yellow dwarf because the light it gives out is yellowish.

Astronomers can figure out the temperature of a star from the colour of the light it gives out. They have found that the temperature on the surface of the Sun is about 10,000°F (5,500°C).

INSIDE THE SUN

Inside, the Sun is much hotter. In the core, or centre, the temperature is estimated to rise as high as 27,000,000°F (15,000,000°C). This great heat affects the hydrogen in the core. It brings about nuclear fusion.

In nuclear fusion, the atoms, or smallest particles, of hydrogen fuse, or join together. They form another gas called helium. When this happens, enormous amounts of energy are given out. This energy gradually makes its way up to the surface of the Sun. There it is given off into space, mainly as heat and light. Only about $\frac{1}{2,000,000,000}$ of the Sun's heat and light reach Earth—the rest goes into space.

Above: The Sun is much bigger than Earth and even the biggest planet, Jupiter. But it is tiny compared with the biggest stars, which are called supergiants. Some super-giants are 250 million miles (400 million km) across.

Below: Inside the Sun, atoms of hydrogen join together to form helium atoms. This fusion process produces the energy that makes the Sun shine.

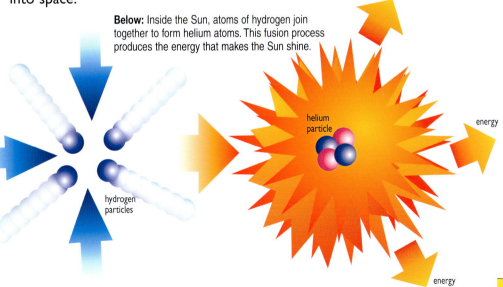

energy

helium particle

energy

hydrogen particles

energy

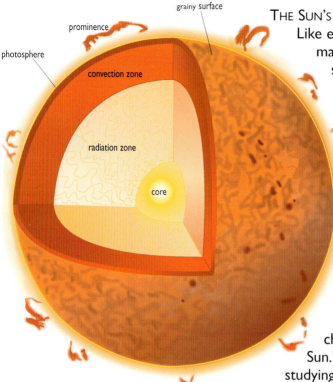

photosphere

prominence

grainy surface

convection zone

radiation zone

core

A look inside the Sun. Heat produced in the core travels as rays inside the radiation zone. In the convection zone, it travels as currents of hot gas.

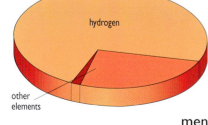

hydrogen

other elements

This pie chart shows the relative amounts of gases in the Sun.

THE SUN'S MAKEUP

Like everything else, the Sun is made up of certain basic substances, called chemical elements. We can think of the chemical elements as the building blocks of matter. On Earth, we find about 90 elements. All the many thousands of different materials on Earth are made up of one or more of these elements.

Scientists have found more than 70 of Earth's chemical elements in the Sun. They have found them by studying the light the Sun gives out. When they pass sunlight through a prism, they get a spectrum, or rainbow of colours. When they look closely at the spectrum, they find that it is crossed by many fine black lines. From the position of these lines, they can tell which chemical elements are present in the Sun.

Hydrogen shows up very clearly. This is because it is the main substance found in the Sun. About ¾ of the Sun is hydrogen, and about ¼ of the Sun is helium.

Carbon, iron, magnesium, lead, silver, and gold are among the many other elements present in the Sun in small amounts. On Earth, these elements are solid materials. But in the intensely hot Sun, they are gases.

THE SUN'S PULL

Every body in space attracts, or pulls, other bodies near it. Earth, for example, pulls us and keeps our feet on the ground. This pull is called gravity. The bigger and heavier a body is, the more powerful is its gravity.

Because the Sun is so big, its gravity is extremely powerful. It is 30 times as strong as Earth's gravity. If you could stand on the Sun's surface, you would not be able to lift your feet. The Sun's powerful gravity keeps the planets in their places in the solar system. It reaches far out into space, to distances of many billions of miles. Even from billions of miles away, it can attract icy lumps a few miles across. When these lumps travel toward the Sun, we see them as comets in the night sky.

THE SUN DATA

Diameter: 865,000 miles (1,392,000 km)

Distance from Earth: 93 million miles (150 million km)

Temperature of surface: 10,000° F (5,500° C)

Temperature of centre: about 27,000,000° F (15,000,000° C)

Nine planets circle around the Sun in space. They are the main members of the Sun's family, or solar system. All the planets lie far apart. The outermost planet, Pluto, sometimes wanders as far as 4.5 billion miles (7 billion km) from the Sun.

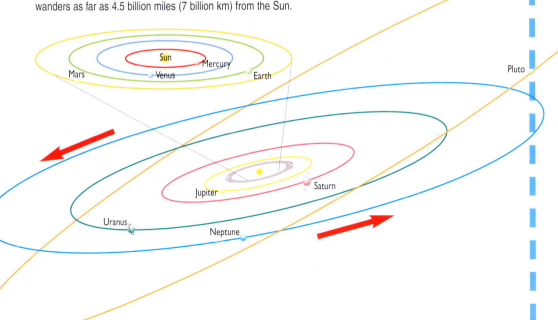

Sun
Mercury
Venus
Earth
Mars
Jupiter
Saturn
Uranus
Neptune
Pluto

Surface and Atmosphere

The bright surface of the Sun boils and bubbles as hot gas rises. Jets of fiery gas shoot up hundreds of thousands of miles into the Sun's atmosphere.

The part of the Sun we see is called the photosphere, which means "light ball." It forms a thin shell around the Sun, about 300 miles (500 km) thick. Most of the heat and light the Sun gives out comes from the photosphere.

The photosphere is constantly heaving like a stormy sea as pockets of hot gas bubble up to the surface. These bubbles make the surface look speckled.

SPOTS ON THE SUN

From time to time, dark patches appear on the photosphere. They are called sunspots. They may grow to be much bigger than Earth and may last for months at a time. Sunspots look darker than the rest of the Sun's surface because they are cooler. Astronomers do not know exactly why sunspots appear. But they do know that it has something to do with the Sun's magnetism, or electrical currents mov-

Above: The surface of the Sun, pictured by the space probe SOHO. The small, bright specks all over the surface show where masses of hot, glowing gas reach the surface. The three large bright patches are powerful solar flares.

Right: A group of sunspots on the Sun's surface. They are about 1,800° F (1,000° C) cooler than the rest of the Sun's surface.

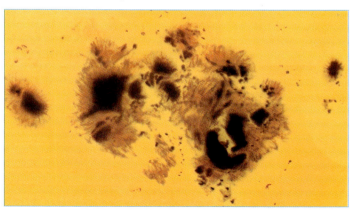

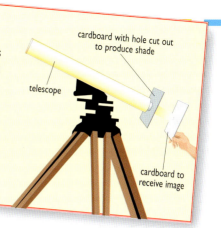

Seeing Sunspots

Using a telescope, you can see sunspots from Earth. But you must not look directly at the Sun because this can seriously damage your eyes. Instead, use the telescope to project an image of the Sun onto a piece of white cardboard. (See diagram at right.) If you watch sunspots from day to day, you will notice that they move across the Sun's surface.

cardboard with hole cut out to produce shade

telescope

cardboard to receive image

A solar flare erupts. A huge sheet of flaming gas springs up in the Sun's lower atmosphere, and streams of particles flow out into space.

ing across the Sun. The Sun's magnetism is very strong where sunspots appear.

From Earth, we can see the sunspots move from day to day. This is because the Sun is spinning around in space. It takes about 25 Earth-days for the Sun to rotate once.

Sunspots come and go on the Sun's surface. In some years, hardly any sunspots appear. In other years, spots spring up all over the place. On average, a year when the Sun has the most sunspots happens every 11 years. Many sunspots were seen in the year 1990.

THE COLOUR BALL

Like Earth, the Sun has an atmosphere made up of layers of gases. The part of the atmosphere just outside the photosphere is called the chromosphere, which means "colour ball." It gets this name because it has a pinkish colour. It is about 6,000 miles (10,000 km) thick.

Like the photosphere, the chromosphere is in constant motion. Little jets of glowing gas dance all over it all of the time. Sometimes huge flickering tongues spring up, like flames in a fire. They are called solar flares.

A solar flare usually lasts for only a few minutes. But during this time, it may become the brightest feature on the Sun. It gives off enormous amounts of energy, and it sends streams of electric particles out into space.

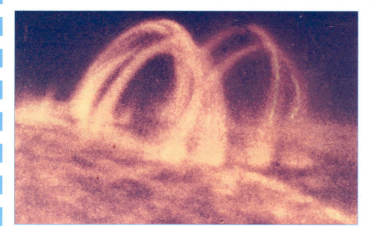

Left: These arches of flaming gas rising high above the Sun are called loop prominences. They form loops when the gas follows the pattern of the Sun's magnetism.

FIERY FOUNTAINS

From time to time, the chromosphere also throws out other jets of flaming gas. They often rise up hundreds of thousands of miles before arching over and plunging back again, somewhat like the water in a fountain. These fiery fountains are called prominences. Prominences often happen close to sunspots. So they are also probably caused by the Sun's magnetism.

THE SUN'S CROWN

Above the chromosphere is the outer layer of the Sun's atmosphere. It is called the corona, which means "crown." It reaches out for millions of miles into space, getting fainter and fainter the farther out it goes.

SEEING THE ATMOSPHERE

The chromosphere and the corona both give off light. But we usually cannot see them because the Sun's surface is so bright. The only time we can see them from Earth is during a total eclipse of the Sun.

Right: This picture from the space probe SOHO shows the bright corona stretching out on each side of the Sun. The face of the Sun has been blotted out (centre) so that the corona becomes visible. In the background are stars of the constellation Sagittarius. At the bottom, part of the Milky Way galaxy can be seen.

Polar Lights

Auroras take place mainly in the far northern and far southern regions of the world, near the North and South Poles. In the north, they are called the aurora borealis, or the Northern Lights. In the south, they are called the aurora australis, or the Southern Lights. Auroras can take many forms. At their most beautiful, they look like shimmering curtains of coloured light.

During a total eclipse, the Moon is positioned between Earth and the Sun and blots out the Sun's light. We can see the chromosphere as a pinkish glow around the edge of the Moon. We can also see prominences, looping up through the chromosphere. Farther out, the corona appears as a pearly white halo.

THE SOLAR WIND

The Sun gives off heat, light, and other rays. It also gives off streams of tiny particles. These particles carry a tiny amount of electricity. They flow out into space in all directions. This flow of particles is called the solar wind.

Usually, the solar wind "blows" gently, like a summer breeze on Earth. But when a flare springs up on the Sun's surface and gives out its great blasts of particles, the solar wind can become a gale.

These particles affect us when they reach Earth. Because they are electric, they can upset our electricity supplies. They can cause interference on the radio. But they also cause beautiful dancing light displays in the sky. These displays are called auroras.

In the Shadows

Sometimes the Moon passes in front of the Sun in the sky and blocks out its light. For a while, day becomes night. The air cools, and birds, thinking it is evening, start to roost.

Earth travels through space with a companion, the Moon. The Moon circles around Earth, while Earth circles around the Sun. With the Sun's light shining on Earth and the Moon, they both make shadows in space, just as you make shadows on the ground when you stand in sunlight.

In July 1991, one of the longest total eclipses of the century took place in Hawaii. Here we see the Moon covering two-thirds of the Sun. It is midmorning, but light is fading fast.

Sun

Moon

partial shadow (penumbra)

complete shadow (umbra)

Earth

A few times every year, the Sun, the Moon, and Earth line up exactly, or almost exactly, in space. When the Moon comes between the Sun and Earth, it makes a shadow that falls on Earth. Someone in the shadow would see the Moon cover all or part of the Sun. This is called an eclipse of the Sun, or a solar eclipse. When the Moon covers all of the Sun, it is called a total eclipse. When the Moon covers only part of the Sun, it is called a partial eclipse.

At other times when the Sun, Earth, and the Moon are lined up, Earth's shadow falls on the Moon. This is called an eclipse of the Moon, or a lunar eclipse. The Moon does not completely disappear from view during a lunar eclipse. It is still lit up by faint light coming around Earth and usually looks red.

WARNING!

You need to look through a special filter to see an eclipse of the Sun. Do not use a piece of over-exposed photographic film or a piece of smoked glass. These will block out the Sun's glare but might let invisible rays through that could damage your eyes.

TOTAL ECLIPSE

During a solar eclipse, the Moon's shadow covers only a small area of Earth's surface. Only people within that area will see a total eclipse. And they see it only for a short period of time. This is because the shadow races across Earth as Earth rotates and the Moon moves overhead.

Total eclipses are among the great spectacles of nature. During a total eclipse, astronomers can study the Sun's atmosphere. They can see the fiery fountains called prominences and the white corona. Astronomers travel all over the world to see and photograph total eclipses.

STARPOINT

The longest a total eclipse of the Sun can last is about 7½ minutes. But most eclipses are much shorter.

The July 1991 total eclipse in Hawaii. Totality, the period of darkness, lasted for more than four minutes. During this time, the Sun's corona shone around the dark Moon. The sky did not go totally dark; it became orange on the horizon, like it often does at sunset.

Frightening the Dragon

Eclipses of the Sun terrified ancient peoples because they did not know what was happening. They thought that some great heavenly monster was trying to swallow the Sun. They knew that if it succeeded they would be doomed, because the Sun brings life to Earth. The ancient Chinese pictured the monster as a dragon. When they saw it starting to swallow the Sun, they banged gongs and cymbals and made as much noise as they could to frighten it away. They found this always worked.

The Sun and the Universe

For us, the Sun is the most important heavenly body there is. But in the Universe as a whole, it is not very important at all. There are billions upon billions of stars like it.

The Universe includes everything that exists. Long ago, people believed that Earth was the centre of the Universe. They thought that the Sun, the planets, and the stars circled around the Earth.

In the 1500s, astronomers realized that Earth was a planet that circled around the Sun. Then they thought that the Sun was the centre of the Universe.

By the beginning of the 20th century, astronomers knew that the Sun was just one of billions of stars in a star system they called the Galaxy. And they thought that the Galaxy was the entire Universe.

1. A massive explosion takes place on the Sun.

2. Some of the millions of stars that travel through space with the Sun

3. The Sun belongs to a spiral galaxy like this.

4. A collection of distant galaxies spotted by the Hubble Space Telescope

These pictures show, from top to bottom, how the Sun fits into the Universe.

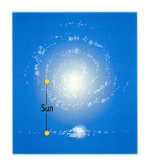

Our galaxy, the Milky Way, is a spiral galaxy, with the stars on arms that curve out from the centre. It is so large that a beam of light would take 100,000 years to travel from one side to the other. Astronomers say it measures 100,000 light-years across.

But soon astronomers began spying other star systems in space. And they realized that our galaxy was only one of many galaxies in space. All these galaxies made up the Universe.

Astronomers estimate that there are more than 15 billion galaxies in the Universe. These galaxies are very far away from our galaxy and from one another. Between them is only empty space. So the Sun is one of many billions of stars in a galaxy, which is one of billions of galaxies that make up the Universe.

THE SUN IN OUR GALAXY

Our galaxy is called the Milky Way. It contains about 100 billion stars. The stars are clustered together into a flat disk, somewhat like a Frisbee. There is a bulge of stars in the centre. The stars in the flat part make up arms that curve out from the bulge. The Sun is found on one of these arms, quite a long way from the Milky Way's centre.

The whole Milky Way is turning around in space. From a distance it would look somewhat like a spinning firecracker. The Sun takes about 225 million years to make one trip around the centre of the Milky Way. This period of time is called a cosmic year.

Light-Years Away

The distances between the Sun and the stars and the galaxies are vast. They are impossible to imagine. For example, the nearest star to the Sun is called Proxima Centauri. It lies more than 25 trillion miles (40 trillion km) away. And this is only a short step in space!

The "light-year," or the distance light travels in a year, is a very handy unit for measuring distances in space. Proxima Centauri is so far away that its light takes over 4 years to reach us. We can say that it lies over 4 light-years away. The light from other stars even farther away can take thousands of years to reach us. In other words, they lie thousands of light-years away.

The brightest stars in this picture lie about 40 light-years away from Earth.

When the Sun Dies

In about 5 billion years, the Sun will start to die. First it will get bigger until it is giant sized. Then it will slowly shrink again until it is a dwarf star about the size of Earth.

Astronomers believe that the Sun is about 4.6 billion years old. Like all stars, it was born from a huge nebula, or cloud of gas and dust. The nebula gradually shrank into a ball of denser, or heavier, matter. The ball got hotter and hotter. Eventually, a nuclear "furnace" lit up inside the ball, which started to shine as the Sun.

The Sun has been shining steadily since then, staying as bright as it is today. But all the while, it has been burning up the hydrogen it uses as fuel. In about 5 billion years, all its hydrogen will be used up. Then the Sun will begin to die.

An enormous red Sun towers in Earth's sky in about 5 billion years time. Millions of years earlier it ran out of hydrogen fuel and began to die. It has since become a star astronomers call a red giant. It will start to shrink in size, finally ending up as a body only the size of Earth.

First, it will start using other fuels to produce its energy. But this will make it swell up. It will get bigger and bigger and redder and redder and turn into a huge body called a red giant. Many of the stars we see in the night sky are red giants.

The Sun might swell in size a hundred times or more until it is more than 90 million miles (145 million km) across. This means that it will then be bigger across than the orbit of the planet Mercury. So Mercury will disappear. The next planet out, Venus, will become scorching hot. Earth will get very hot too, and any life on it will perish.

FROM GIANT TO DWARF

In time, the giant Sun will use up all its fuel. Its nuclear furnace will go out. Then it will start to collapse. As it shrinks, it will puff off some of its gas from time to time. This will form a colourful cloud around it, which will gradually grow bigger.

The Sun, though, will continue to get smaller and whiter. In time, it will shrink into a body called a white dwarf. The white dwarf Sun will be about the same size as Earth. But it will be very much heavier. A table-spoonful of its matter might weigh as much as a thousand tons!

Gradually, the white dwarf Sun will cool down and fade. It will become dimmer and dimmer until it stops shining altogether. Then it will die, becoming an invisible black speck in the Universe.

nebula

Sun is born

The life and death of the Sun. It was born in a cloud of gas and dust. It will swell up when it starts to die, then shrink again. Finally it will fade away.

the Sun in A.D. 2,000

red giant

planetary nebula

white dwarf

Blowing Rings

Astronomers know of other stars that were once like the Sun but are now dying. They can see the nebulae that the stars have puffed off at the end of their lives. The most famous example is called the Ring nebula because it looks like a smoke ring. In the middle of the ring is the star that is dying. This kind of nebula is called a planetary nebula. In small telescopes, it appears round, like a planet.

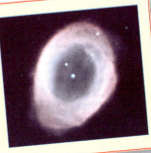

Exploring the Sun

This tall tower is one of the solar telescopes at the Mount Wilson Observatory, near Los Angeles. A mirror on top reflects an image of the Sun down the shaft in the centre to the observing room.

Astronomers study the Sun in many ways. They use special telescopes at observatories on the ground. They also use satellites in orbit around Earth and probes that travel deep into space.

Because the Sun is so important to us, we need to know as much about it as possible. Changes that take place on and in the Sun can affect Earth. For example, solar flares can cause electric storms here. Changes in the Sun's heat output can cause changes in our climate.

Astronomers build special telescopes to study the Sun. Solar telescopes take the form of tall towers. They have mirrors on top to reflect sunlight down to an observation room at the bottom. In this room, astronomers study images of the Sun's surface.

Kitt Peak Observatory in Arizona has the biggest solar telescope in the world. It is called the McMath Solar Telescope. It produces an image of the Sun's surface more than 3 feet (1 m) across.

The tower of the telescope stands about 11 stories high. Its mirror reflects sunlight down a sloping tunnel cut into the mountainside. A second mirror at the bottom reflects the light to a third mirror at ground level. This mirror reflects an image of the Sun into the observing room.

The McMath Solar Telescope at Kitt Peak Observatory, near Tucson, Arizona. A mirror on top of the upright tower reflects sunlight down the sloping shaft and deep underground.

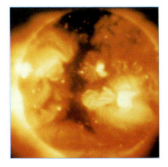

Above: The Sun as seen through Skylab's X-ray telescope. The dark areas are holes in the Sun's outer atmosphere, or corona.

OBSERVING FROM SPACE

Astronomers cannot get a complete picture of what the Sun is like using solar telescopes on Earth. This is because Earth's atmosphere blocks many of the invisible rays coming from the Sun. For example, it blocks X rays and many ultraviolet rays.

This is one reason astronomers send solar telescopes into space. In space they can study all the rays coming from the Sun. This gives them a much more complete picture of what the Sun is like.

Three teams of astronauts on the experimental U.S. space station Skylab carried out the first main study of the Sun from space in 1973 and 1974. They took tens of thousands of pictures of the Sun with ultraviolet light and X rays, as well as with ordinary light.

Space station Skylab in orbit in 1973. It observed the Sun with eight solar telescopes, mounted in the circular dish you see in the centre of the picture. The X-shaped panels carry solar cells to make electricity. On the right, you can see the patches astronauts used to repair Skylab after it had been damaged during launching.

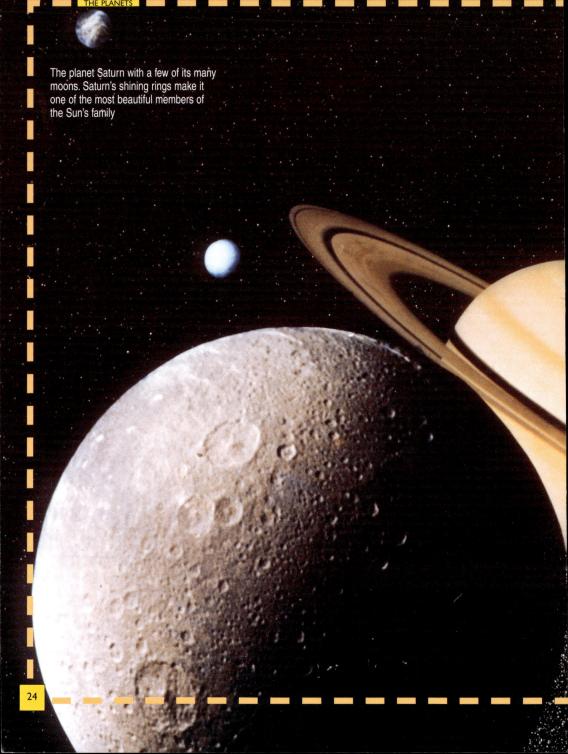

The planet Saturn with a few of its many moons. Saturn's shining rings make it one of the most beautiful members of the Sun's family

The Solar System

Earth, the planet we live on, belongs to a family of bodies that travel together through space. Earth is part of the Sun's family, which is called the solar system. The word solar means "having to do with the Sun." The Sun is a star—a huge ball of glowing gas.

Earth is one of nine large bodies (if we include tiny Pluto) that circle in space around the Sun. We call these bodies the planets. Many other bodies belong to the solar system. Smaller bodies known as moons circle around most of the planets. Smaller still are swarms of rocky lumps we call asteroids. They circle in a huge ring between the planets Mars and Jupiter. In addition, smaller particles known as meteoroids hurtle through space around the Sun.

Some of the most fascinating and least-known members of the solar system are the dusty "snowballs" we call comets. They visit our skies from time to time, often growing spectacular tails.

The newest members of the solar system are spacecraft that we have launched into space. These include satellites that circle Earth, and space probes that journey to the Moon and to distant planets, asteroids, and comets.

It is thanks to these spacecraft that we know so much about our solar system. These craft have also shown us that solar systems exist around other stars in the heavens. So there are probably many other planets in our Universe. Some of these other planets could possibly be like our home planet, Earth.

In the Beginning

The solar system was born from a huge cloud of gas and dust. The Sun, the planets, and other bodies formed when the cloud collapsed under gravity.

The time is nearly 5 billion years ago. A huge cloud of gas and dust in space begins shrinking (1) and starts to spin around (2). As it shrinks more and more, it starts to flatten out (3). Inside, it is warming up.

Long before the solar system was formed, there was nothing in our corner of space but a great cloud of gas and dust. Many clouds like this exist throughout space. Astronomers call them nebulae. It is in nebulae that stars are born. The nebula that once occupied our corner of space was the birthplace of the star we know as the Sun.

This nebula started to shrink, or collapse, about 4.6 billion years ago. The collapse happened because of gravity. Gravity is the pull, or attraction, that every bit of matter has on objects on or near it. Specks of dust and wisps of gas in the nebula began to attract one another and to form a denser, or more tightly packed, mass.

SPINNING AROUND

As the cloud started to shrink, it also started to spin around. Over time, it turned into a thick disk with a large bulge in the centre. At this stage, it would have looked something like a fat spinning Frisbee.

The bulge at the centre kept on shrinking. As it became denser, it began to heat up. The heat came from the energy in the colliding particles. Now looking more like a ball, the bulge got smaller and smaller, and hotter and hotter.

The disk of matter around the hot ball gradually flattened out. Small lumps of rock and metal filled the warm inner parts of the disk. In the cold outer parts, there were lumps of ice and freezing gases.

Birthplace of Stars

Stars galore are being born in nebulae all the time. From Earth, we can easily see one of these nebulae with the naked eye in the constellation, or star pattern, we call Orion. This constellation stands out in the night sky in winter. The Orion nebula lies near the line of three stars in the middle of the constellation. The nebula glows brightly because it reflects light from young hot stars within it.

The Orion nebula, photographed by the Hubble Space Telescope

THE SUN SHINES

The hot ball in the centre of the spinning disk continued to shrink until it was about 900,000 miles (1,400,000 km) across. By this time, it had become extremely hot.

Inside, its temperature rose as high as 27,000,000° F (15,000,000° C).

At this temperature, atoms of hydrogen gas in the ball began to combine, or fuse together. Fantastic amounts of energy were produced in this process. The ball gave off this energy as light and heat. It began shining as a star—the star we call the Sun.

THE PLANETS FORM

Little lumps of material were whizzing around in the spinning disk surrounding the new Sun. They were continually bumping into one another and sticking together to form larger and larger lumps. Gravity was again at work. Over millions of years, the lumps grew and eventually formed the planets and their moons.

Four planets—Mercury, Venus, Earth, and Mars—formed in the warmer, inner part of the disk. They formed from the heavier lumps of rock and metal found there. Five planets formed in the colder, outer part of the disk—Jupiter, Saturn, Uranus, Neptune, and Pluto. They formed from the lumps of ice and freezing gas found there.

4

Above: The planet Mars, photographed by the Hubble Space Telescope. It is one of the four rocky planets, which formed in the warmer, inner regions of the solar system.

5

The rotating cloud of gas and dust continues to flatten into a disk (4). At its centre, a ball of matter grows, heats up, and starts to glow. Around it, bits of matter start to lump together (5).

Over time, the ball in the centre starts shining as the Sun, and the circling lumps turn into planets. The solar system as we know it has been formed (6).

6

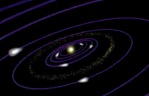

THE LEFTOVERS

Many smaller lumps of matter were left over after the planets and their moons had formed. They were scattered throughout the space between the planets. Some of these lumps gathered together in the space between Mars and Jupiter to form a huge ring of material. We call them the asteroids.

Some lumps, known as meteoroids, rained down on the newly formed planets. They dug out deep holes, or craters, in the planets' surfaces. Meteoroids still circle the Sun. Other lumps of matter remained on the edges of the solar system. We see them only when they wander in toward the Sun and start to shine as comets.

IN THE END

The solar system will probably stay much the same as it is for another 5 billion years. Then the Sun will start to die. It will probably grow bigger and bigger until it stretches out as far as the planet Venus. The Earth will be baked by the Sun's heat. The solar system as we know it will come to an end. Astronomers predict this ending for the solar system because they have seen the same thing happen to other stars like the Sun.

Below: The planet Neptune, photographed by the Hubble Space Telescope. It is one of the four giant planets, made up mainly of gases, which formed in the colder, outer regions of the solar system.

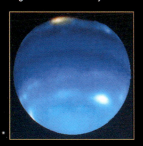

Mapping the Solar System

The solar system stretches over a vast region of space. In a space shuttle, it would take you more than 100 years to travel from one side to the other.

The Sun lies at the heart of the solar system. The nine planets that circle around it form the main part of the Sun's family. They circle around the Sun at different distances. The closest planet to the Sun is Mercury. Then, in order going out from the Sun, are Venus, Earth, Mars, Jupiter, Saturn, Uranus, Neptune, and Pluto.

PLANET ORBITS

Each planet travels around the Sun in a path we call its orbit. The orbit never changes. Most of the planets travel in orbits that are nearly circles. Others travel in orbits that are more oval, or elliptical, in shape. Mercury and Pluto have very elliptical orbits. The Sun

Below: The orbits of the four inner planets—Mercury, Venus, Earth, and Mars

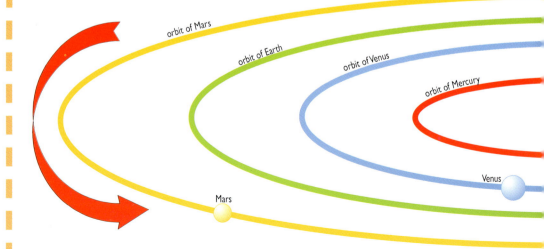

orbit of Mars

orbit of Earth

orbit of Venus

orbit of Mercury

Venus

Mars

STARPOINT

Without the Sun's gravity to hold them in orbit, the planets would fly off into space.

keeps the planets in their orbits with its enormous gravity. This pull is so powerful that it acts over distances of many billions of miles.

Mercury's average distance from the Sun is about 36 million miles (58 million km). At its farthest distance, Pluto travels more than 5 billion miles (7 million km) away. Comets and other small bodies lie at even greater distances. All the bodies in the solar system are far apart. In fact, most of the solar system consists of empty space.

THE INNER PLANETS

The four planets nearest the Sun, from Mercury to Mars, lie close together compared with the other planets. They form a small family of their own—the inner planets. Earth is the largest of the inner planets. The other three inner planets are also rocky bodies like Earth. They are often called the terrestrial, or Earth-like, planets.

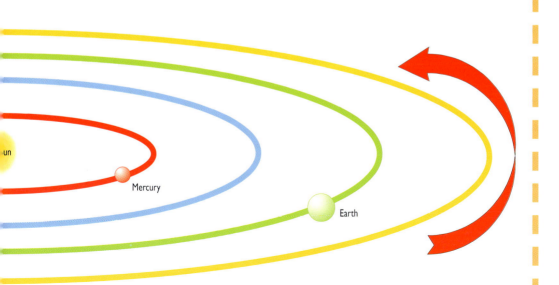

Sun

Mercury

Earth

Astronomical Units

The solar system is very large. From one side to the other it measures more than 10 billion miles (16 billion km). It is difficult for anyone to imagine distances so large. So astronomers often measure distances in astronomical units (AU). One astronomical unit is the distance between the Sun and Earth, about 93 million miles (150 million km). Using this unit, distances become easier to understand. Neptune, for example, is 30 times farther away from the Sun than Earth is.

Planet	Distance from Sun (AU)	Planet	Distance from Sun (AU)
Mercury	0.4	Saturn	9.5
Venus	0.7	Uranus	19.0
Earth	1.0	Neptune	30.0
Mars	1.5	Pluto	39.0
Jupiter	5.2		

THE OUTER PLANETS

The inner planets make up only a tiny part of the solar system, as the drawing at the bottom of the page shows. The remaining planets—Jupiter, Saturn, Uranus, Neptune, and Pluto—take up a far larger part of the solar

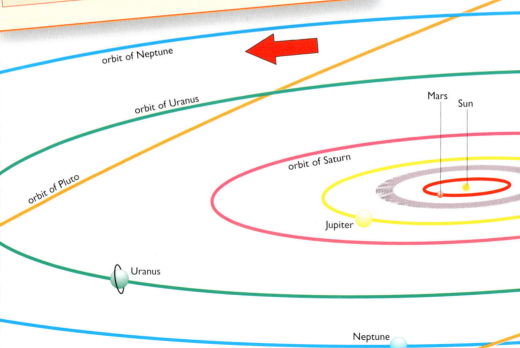

orbit of Neptune

orbit of Uranus

orbit of Pluto

orbit of Saturn

Mars   Sun

Jupiter

Uranus

Neptune

system. We call them the outer planets. They lie much farther from the Sun, and from one another, than the inner planets do.

SAME PLANE

The drawing below shows another feature of the solar system. Most of the planets travel in the same plane. This means that if you could place the Sun and Earth on a flat sheet, the other planets would be on the sheet too. The odd planet out is Pluto. Pluto travels in an orbit that takes it far above and below the plane in which the other planets travel.

SAME DIRECTION

All the planets also travel in the same direction in their orbits around the Sun. If we could look down on the solar system from a point in space high above Earth's North Pole, we would see the planets travelling counterclockwise.

BEYOND THE PLANETS

Pluto is the most distant of the planets. But it is not the most distant body in the solar system. Much farther out are great clouds of small icy bodies. These clouds may reach out trillions of miles from the Sun. From time to time, some of the icy lumps leave the clouds and journey in toward the Sun. We then see them shine as comets.

orbit of Pluto

orbit of Neptune

orbit of Uranus

asteroid belt

Saturn

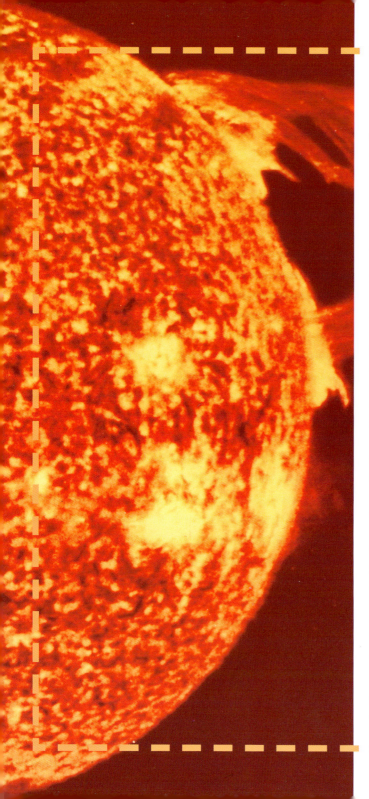

Family Portraits

The many members of the Sun's family are all very different from one another. They vary in size, makeup, and appearance.

STAR OF THE SOLAR SYSTEM
The Sun is quite a different body from all the other members of its family. It is a huge ball of white-hot gases, so big that it could swallow a million bodies the size of Earth. It is a kind of body we call a star. It looks so much bigger and brighter than the other stars in the sky because it is much closer to us.

The Sun is the only body in the solar system that gives off light. The Moon and the planets shine in the night sky, but only because they reflect light from the Sun. If the Sun did not shine, the whole solar system would be in darkness.

This close-up of the Sun's surface shows a boiling, bubbling mass of very hot glowing gas.

COLD PLANETS

There are four other huge balls of gas in the solar system. They are four of the outer planets—Jupiter, Saturn, Uranus, and Neptune. We call them the gas giants. But they are not hot like the Sun. They are very cold and are made up mainly of gas and liquid gas. The other outer planet, tiny Pluto, is very cold too.

...AND HOT PLANETS

The four warmer, inner planets—Mercury, Venus, Earth, and Mars—are much smaller and are made up mainly of rocks. Our home planet, Earth, is the biggest rocky planet. Its rock is cold only near the surface. We see hot rock from below when it forces its way up to the surface in volcanoes.

Above: Saturn is one of the cold gas giants of the outer solar system. At the top of its atmosphere, the temperature is about –300° F (–185° C).

Below: Saturn is one of the cold gas giants of the outer solar system. At the top of its atmosphere, the temperature is about –300° F (–185° C).

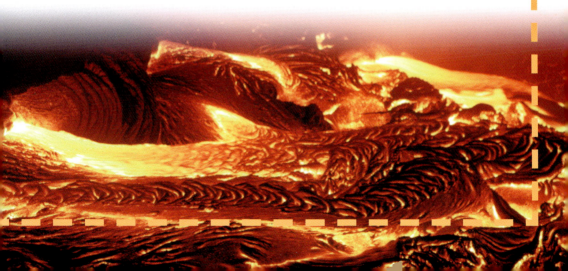

Saturn has 18 known moons —more than any other planet. The largest ones are shown in the picture. In the foreground is Dione. At top right is Saturn's largest moon, Titan, which is bigger than the planets Mercury and Pluto.

This is the minor planet, or asteroid, we call Gaspra, a rocky body only about 10 miles (17 km) long. The space probe Galileo took this photograph when it flew past the asteroid in 1991.

MANY MOONS

Smaller than the planets are the bodies known as moons, which circle most of the planets. Only Mercury and Venus do not have moons. Earth has one moon, which we call the Moon. Mars has two moons. Between them, the four giant planets have at least 59 moons. The two biggest moons in the solar system are Jupiter's Ganymede and Saturn's Titan. They are both bigger than the planets Mercury and Pluto.

Moons are made up of different materials. Our own Moon is made mainly of rock, much like Earth. The large moons of the giant planets are a mixture of rock and ice. Jupiter's moon Io is unusual because it has volcanoes erupting all over it. No other moon has volcanoes.

MINOR PLANETS

Between the orbits of Mars and Jupiter is a ring of small rocky bodies called asteroids. They are also known as the minor planets. Even the biggest, called Ceres, is only about 600 miles (1,000 km) across. Astronomers once thought that these bodies were the remains of another planet that broke up into pieces long ago.

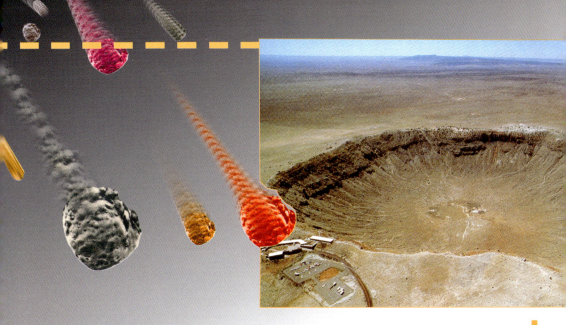

FALLING STARS

Sometimes in the night sky you can see little streaks of light. It looks as if some of the stars are falling down. The stars are not falling, of course. What you see are meteors. These are fiery streaks made by specks of rock called meteoroids. These specks hit Earth's atmosphere travelling very fast. Friction from the air heats the meteoroids. They glow red-hot and then burn up, leaving behind a fiery trail.

Some of the bigger meteoroids do not burn up completely. They fall to planets and moons as meteorites. Large ones dig out deep piys, or craters, in the surface.

HAIRY STARS

Comets are some of the smallest members of the Sun's family but are among the most spectacular to look at in the night sky. Just a few miles across, these icy lumps visit our skies after travelling for billions of miles from the outer parts of the solar system.

As they near the Sun, comets give off great shining clouds of gas and dust. These clouds often fan out from the comet to form a tail millions of miles long.

Above: Thousands of years ago, a meteorite landed in Arizona and created the famous Barringer Crater.

Below: The best known of all "hairy stars," or comets, is Halley's Comet. It last appeared in our skies in 1986 and will not be seen again until the year 2061.

Wandering Stars

The planets shine in the night sky like bright stars. But they change their position night by night against the background of true stars.

If you look at the night sky at about the same time for several nights, you will see the same patterns, or constellations, of stars. Ancient astronomers called the stars in the constellations the fixed stars. They called the planets wandering stars. The planets looked like stars but were constantly on the move among the never-changing constellations. The word planet means "wanderer."

You can see another difference between planets and stars. Stars twinkle, but planets do not. Air currents in the atmosphere make the faint light from the distant stars wobble. This causes them to twinkle. The stronger light from the planets is not affected, so they shine steadily.

BRIGHT AND BEAUTIFUL
From Earth we can see five planets without a telescope. Venus is the brightest by far. It shines brighter than any star. Both Venus and Mercury can sometimes be seen

In the night sky, the planets can be found within an imaginary band called the zodiac. The zodiac contains several star patterns, or constellations. Ancient astronomers named the constellations after figures they imagined they could see in the patterns of stars. We still use these names for the constellations of the zodiac.

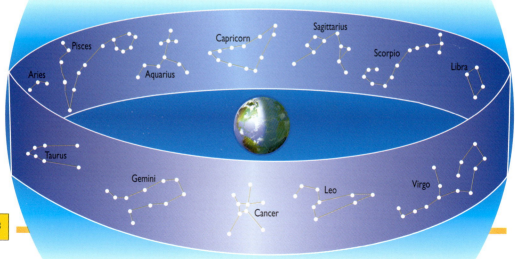

just before sunset or just after sunrise. Mars, Jupiter, and Saturn also shine brightly in the night sky. They are easy to tell apart. Mars shines with a reddish-orange colour. It is often called the Red Planet. Jupiter appears to be a brilliant white, and Saturn's light is a yellowish white.

THE CIRCLE OF ANIMALS

The planets do not wander everywhere in the night sky. They can be found only among certain stars. The band of stars in which they are found is called the zodiac. The word zodiac means "circle of animals." This name comes from the 12 constellations the band contains. Most of these constellations have the names of animals, such as Aries (the Ram) and Taurus (the Bull).

Above: The zodiac constellation Taurus, the Bull. Ancient astronomers imagined they could see in this star pattern the head of a charging bull.

Written in the Stars

In ancient times, most people believed that the heavenly bodies somehow affected human lives. This idea is called astrology. Some people still believe in astrology today. Astrologers study the positions of the Sun and the planets in the constellations of the zodiac. They believe they can use this knowledge to predict events in people's lives.

Sagittarius
Libra
Aries
Taurus
Gemini
Cancer

Leo
Virgo
Scorpio
Capricorn
Aquarius
Pisces

Comparing the Planets

The planets vary in all sorts of ways. Some are big, others small. Some are rocky, others are made of gas. Some spin slowly, others quickly. Some circle the Sun in days, others take years.

Our home planet, Earth, seems like a big place. But it is tiny compared with some of the other planets. The widest part of Earth is at the equator. If you drove a car around the equator at a speed of 50 miles (80 km) an hour, it would take you three weeks to circle Earth. But it would take you more than six years to drive a car around Jupiter!

We can compare the planets' sizes with the drawings on these pages. Jupiter is by far the biggest planet. Earth is the middle-sized planet. Four planets are bigger than Earth and four are smaller. Tiny Pluto is even smaller than Earth's Moon.

Jupiter may be huge for a planet, but it is tiny compared with the Sun. If the Sun were the size of a beach ball, Jupiter would be the size of a golf ball. Earth would be smaller than an orange seed!

Some of the moons in the solar system are bigger than some of the planets! Saturn's moon Titan is bigger than Mercury and Pluto. It measures 3,193 miles (5,140 km) across.

This picture shows all the planets drawn to the same scale. It shows just how small Earth and the other inner planets are compared with Jupiter and the other gas giants.

Mercury

Venus

Earth

Mars

Jupiter

	Diameter at equator (miles)	Average distance from Sun (million miles)	Rotates in:	Orbits Sun in:	No. of moons
Mercury	3,031	36	50 days	88 days	0
Venus	7,521	67	243 days	225 days	0
Earth	7,926	93	24 hrs	365 days	1
Mars	4,222	142	24.6 hrs	687 days	2
Jupiter	88,400	484	9.9 hrs	12 yrs	16+
Saturn	74,600	687	10.6 hrs	30 yrs	23+
Uranus	31,800	1,783	17.2 hrs	84 yrs	15+
Neptune	30,800	2,795	16 hrs	165 yrs	8
Pluto	1,420	3,670	6.3 days	248 yrs	1

Among the gas giants, Saturn is slightly smaller than Jupiter. Uranus and Neptune are nearly the same size. Tiny Pluto is by far the smallest planet, smaller even than Earth's Moon.

Saturn

Uranus

Neptune

Pluto

Right: A look inside Neptune, an example of a gas giant. It is made up of different layers, beginning with an atmosphere thousands of miles deep. We cannot see through this thick outer atmosphere. Underneath it is an even deeper ocean. And at the centre is a rocky core.

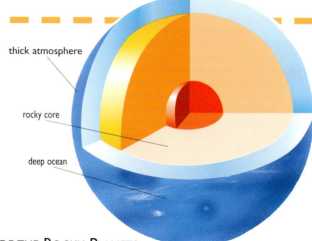

thick atmosphere

rocky core

deep ocean

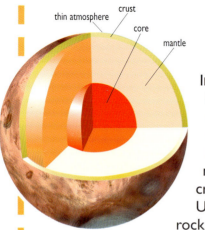

thin atmosphere   crust

core

mantle

A look inside Mars, an example of a rocky planet. Like Neptune, Mars is also made up of different layers. But it has only a very thin atmosphere, which we can see through. Underneath is a hard crust, which lies on top of the mantle. At the centre is the core, which is probably solid, unlike Earth's liquid core.

INSIDE THE ROCKY PLANETS

In their structure, or makeup, planets fall into two main groups—rocky planets and gas giants. Earth is a typical rocky planet. At the surface is a hard outer layer of rock, called the crust. Earth is unique in that much of the crust is covered by water. Above the crust, a thin layer of gases forms the atmosphere.

Underneath Earth's crust is a thick layer of softer rock, called the mantle. This layer gets hotter the farther down you go. At Earth's centre is a mass of metal, mainly iron. This is called the core. The temperature of the core is so high that the metal is liquid.

The planets Mercury, Venus, and Mars are made up of layers of rock in much the same way. That is why they are called the terrestrial, or Earth-like, planets.

INSIDE THE GAS GIANTS

The four giant planets, Jupiter, Saturn, Uranus, and Neptune, are quite different in makeup from the terrestrial planets. They are made up mainly of gases and cold liquid gases.

Jupiter is a typical gas giant. It has a very deep atmosphere of gases, mostly hydrogen gas. Underneath the atmosphere is a deep ocean of cold liquid gas. There is no solid surface at all. At the centre of the planet is a small core of rock.

The Earth spins in space on an imaginary line called its axis. The line goes through the North and South Poles. All the planets spin around on an axis in a similar way.

axis

North Pole

South Pole

PLANET MOVEMENTS

All the planets move in two ways. Each planet travels in its orbit around the Sun. Earth takes a little over 365 days, or one year, to travel once around the Sun. Planets closer to the Sun take a shorter time to make one orbit because they have a shorter distance to travel. Planets farther from the Sun than Earth take a longer time because they have a longer distance to travel.

Each planet also spins around on its axis. A planet's axis is an imaginary line that runs through it, around which it spins, or rotates. Earth takes 24 hours to spin around once. This is the period of time we call one day. Some planets spin faster than Earth; some spin more slowly. The table on page 41 shows how fast the planets spin and how long they take to orbit the Sun.

The Life Zone

There is one big difference between Earth and all the other planets. Earth is covered with living things. There are millions of different kinds of plants and animals, from tiny weeds to monster whales. Life does not exist on any other planet. One main reason for life on Earth is its position in the solar system. Its distance from the Sun means that it is not too hot and not too cold for life. We say that Earth lies in the Sun's life zone. Earth's neighbours in space, Venus and Mars, lie just outside the life zone. Venus is too hot, and Mars is too cold for life to exist.

Exploring the Solar System

The Hubble Space Telescope, photographed in orbit by astronauts in the space shuttle. Astronauts travel to the telescope every few years to check its instruments and other parts, such as the solar panels. These gold-coloured panels on each side power the telescope. The masts sticking out from the telescope body carry radio antennae. These antennae receive signals from mission controllers and also send back the pictures the telescope takes.

Through the camera eyes of space satellites and probes, astronomers can take a close look at the planets and their moons, as well as asteroids and comets.

The ancient astronomers had no idea what the planets and the other bodies in the solar system were like. They knew of only five planets—Mercury, Venus, Mars, Jupiter, and Saturn. These planets appeared to them as bright shining stars.

In the 1600s, astronomers such as Galileo began using telescopes to look at the night sky. They discovered many new things about the solar system. For example, they learned that the planet Jupiter had moons circling around it.

Later, astronomers using more powerful telescopes began discovering new planets: Uranus (in 1781), Neptune (in 1846), and Pluto (in 1930). But even with the big telescopes available to modern astronomers we still cannot see many details on any of the planets. They are too far away.

ROBOT EXPLORERS

Astronomers therefore use other ways of exploring the solar system. One way is by using robots. These

The Hubble Space Telescope took this picture of Jupiter. It clearly shows the colourful bands of clouds in the atmosphere. It also shows Io, one of Jupiter's moons. The dark spot is Io's shadow.

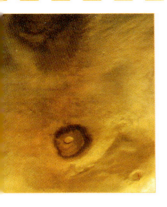

The surface of Mars, photographed by the space probe Mariner 9 in July 1965. It was the first time that a probe had photographed a planet up close.

robots are actually spacecraft that carry telescopes, cameras, and other instruments.

Telescopes on spacecraft get a much clearer view of the heavenly bodies than telescopes on Earth. Telescopes on Earth peer at the heavens through the atmosphere, which is full of dust and moisture that spoil the view. But spacecraft travel high above the atmosphere and are not affected by it.

ASTRONOMY SATELLITES

Two main kinds of spacecraft are used to explore space. One is the astronomy satellite, which stays in orbit around Earth and looks at the heavenly bodies from a distance.

The Hubble Space Telescope is an example of an astronomy satellite. It has been in orbit since 1990. And it is sending back amazing pictures of planets, comets, asteroids, nebulae, and stars. The telescope is a huge instrument, measuring 43 feet (13.1 m) long and weighing more than 11 tons. It circles in space about 320 miles (515 km) above Earth. It picks up light from the heavenly bodies with a curved mirror 95 inches (2.4 m) across.

The First Encounter

The first spacecraft to travel close to another planet was Mariner 4. It set out for Mars in November 1964 and reached the planet in July 1965. It flew past Mars at a distance of about 6,000 miles (9,700 km). It took about 20 pictures, which showed a number of craters on the surface. Instruments on Mariner 4 also found that Mars had a thin atmosphere, made up mainly of carbon dioxide. The last Mariner spacecraft to orbit Mars was Mariner 9, in 1971. It made a map of the whole planet.

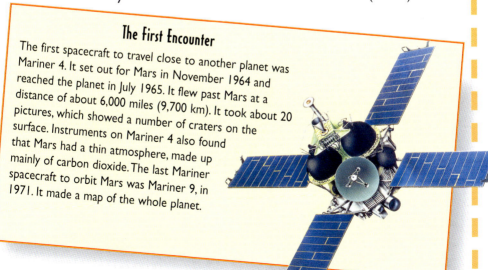

August 20, 1977: the launch of Voyager 2.

PROBES TO THE PLANETS

Astronomers also use a spacecraft called a space probe to explore the solar system. Space probes are unmanned vehicles that leave Earth behind and travel into the depths of space. Probes are sent to the Moon, to the planets and their moons, and to comets and asteroids.

The first successful probes to the planets were sent to Mars and Venus in the 1960s. Since then, probes have visited all the planets except Pluto. They have sent back amazing information and pictures. They have shown us that Mercury looks much like the Moon; that Jupiter's moon Io has volcanoes; that winds on Saturn can blow at more than 1,000 miles (1,600 km) an hour; and that Neptune's moon Triton has geysers.

FAST MOVERS

Very powerful rockets are needed to launch a probe to a target that may be billions of miles away. A rocket must make a probe go fast enough to escape the pull of Earth's gravity. To escape, a probe has to be launched at a speed of 7 miles (11 km) per second. At this speed, you could fly across the Atlantic Ocean in about 8 minutes!

Jupiter:
July 9, 1979

Voyager 2 set off on its historic journey to the outer planets atop a Titan-Centaur rocket in August 1977. Less than two years later, it was flying past the giant planet Jupiter. Boosted by Jupiter's gravity, it sped on to Saturn, then to Uranus, and finally to Neptune. It arrived at Neptune after spending 12 years travelling through space.

Saturn:
August 25, 1981

Triton, Neptune's largest moon, was Voyager 2's last port of call. This icy moon is covered with a pinkish snow of frozen gases. Dark material thrown out from icy volcanoes is visible in places.

FLY-BY, ORBIT, OR LANDING

Most probes carry out a fly-by, which means they fly past a planet or other body. The Voyager probes to the outer planets and the Giotto probe to Halley's comet carried out fly-bys. Some probes fly to a planet and then go into orbit around it. The Magellan probe to Venus and the Galileo probe to Jupiter were orbiters.

Other probes travel to a planet or a moon and drop landing craft down to the surface. The Viking probes to Mars did this. The Pathfinder probe that landed on Mars carried a wheeled vehicle down to the surface. This rover, called Sojourner, travelled around investigating nearby rocks. It was guided by scientists back on Earth.

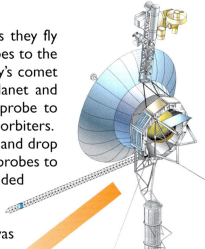

Voyager 2

Uranus:
January 24, 1986

Neptune:
August 24, 1989

Other Solar Systems

Above: Astronomers have found a disk of matter around a star named Beta Pictoris. Over time, the matter in the disk could lump together to form planets.

Right: This is part of the famous Orion nebula, as seen by the Hubble Space Telescope. It shows a number of small disks where solar systems are forming. Before the Hubble pictures, astronomers had never seen such disks clearly. They call these newly born solar systems proplyds (short for protoplanetary disks).

Out in space, in other solar systems, there may be planets like Earth. These planets might swarm with life like Earth does.

The Sun was born in one of the great clouds of gas and dust that exist in space. As far as we know, all stars form in the same way. This means that other stars could form some kind of solar system, containing planets like those in our own solar system.

Some astronomers think they can detect a few stars with planets. They notice that these stars wobble slightly, as though they are being pulled by unseen bodies, which could be other planets. The planets themselves would be too small to be seen even in the most powerful telescopes on Earth.

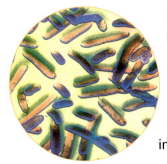

Even using space telescopes, we cannot see individual planets near distant stars. But we are able to spot disks of matter around some of the stars. The Hubble Space Telescope has sent back spectacular pictures of such disks in the Orion nebula. One day, the gas and dust in these disks could lump together to form planets, which is what happened in our own solar system long, long ago.

Bacteria like this one, E. coli, are found everywhere on Earth, in the air, in the ground, and in other living things. Scientists think that similar kinds of organisms were among the first forms of life on Earth.

PLANETS LIKE EARTH

In our Universe, there are many stars like the Sun. And many of these stars are probably the centre of a solar system. And in many of these solar systems, there are probably planets like Earth. Because there are billions of stars like the Sun, there could be billions of planets like Earth.

Many of these planets might circle in the life zones of stars, where the temperature would be not too hot and not too cold. So on these planets, life of some kind could exist. There could even be intelligent life-forms like human beings.

How Life Began

Scientists do not know whether life exists elsewhere in space. One reason for this is that they do not really know how life came about on Earth. Many scientists believe that life began with chemicals that formed in Earth's atmosphere billions of years ago. They formed when lightning acted on the gases it contained. The chemicals rained into the oceans and joined to make very simple living things. These early life-forms were like the tiny life forms we know as bacteria. Then over many hundreds of millions of years, more advanced life-forms—plants and animals—developed.

Planet Earth

Earth is the planet on which we live. It is part of the solar system, or family of bodies that travel around the Sun. Earth circles the Sun along with eight other planets. As planets go, Earth is fairly small—its diameter, or distance through Earth from North Pole to South Pole, is less than one-tenth the diameter of Jupiter. But Earth's diameter is more than five times that of the smallest planet, Pluto.

Earth formed at the same time as the other planets, about 4.6 billion years ago. It is made up mainly of different kinds of rock. About 70 percent of Earth's surface is covered by water. Land areas cover the rest. The land areas form the continents, and the water areas form the oceans.

Earth is also surrounded by gases. The gases form a layer around Earth called the atmosphere. One of the gases in the atmosphere is oxygen. This is the gas almost all animals must breathe in order to stay alive.

Plant and animal life thrives on Earth in millions of different forms on the land, in the oceans, and in the air. Conditions on Earth are just right for life. Water and air are available, and temperatures are comfortable because Earth is not too close to and not too far away from the Sun.

As far as we know, Earth is the only planet in our solar system that has living things. Indeed, it is the only place in the universe that we know has living things.

From space, Earth looks blue because of its oceans. This photograph shows the continent of Africa, with the Atlantic Ocean on the left and the Indian Ocean on the right.

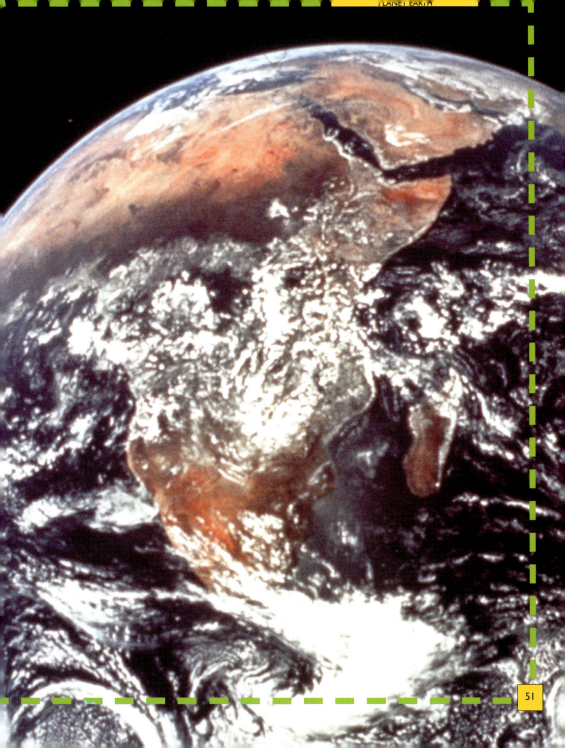

Our Home Planet

Earth is one of the four small, rocky inner planets, that lie close to the Sun. Like all planets, Earth travels around the Sun, and it turns around like a top.

Earth is the third planet out from the Sun, which lies about 93 million miles (150 million km) away. Earth's nearest neighbours among the planets are Venus, Mars, and Mercury. They are made up mainly of rock, like Earth. These four inner planets are called the terrestrial, or Earth-like, planets.

Earth is different from the other rocky planets because most of the rocky surface is covered by the water of the oceans.

SPEEDING THROUGH SPACE

Every day we see the Sun travel slowly across the sky. Every night we see the stars wheel slowly overhead. Earth seems to stand still. But the opposite is true. The Sun and the other stars only seem to move. In fact, they stand still and Earth moves. Earth is turning around in space like a top. It takes one day for Earth to turn, or rotate, all the way around on its axis. An axis is an imaginary line running through a planet from its north pole to its south pole.

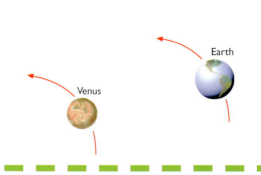

Mars

Earth

Venus

Mercury

Sun

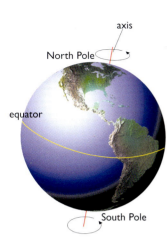

axis

North Pole

equator

South Pole

Earth spins around an imaginary line called the axis. The equator is another imaginary line midway between the North and South Poles.

At the same time that it rotates, Earth also travels through space in a path, or orbit, that takes it around the Sun. It orbits all the way around the Sun in a little over 365 days, a period we call one year.

THE BIG ATTRACTION

Like the other planets, Earth is spherical, or shaped like a ball. But it is not perfectly round. It bulges a little around the equator, which is an imaginary line around Earth, midway between the North and South Poles.

Earth weighs more than 6,000,000,000,000,000,000,000 tons. Like every other massive object, Earth has a powerful attraction, or pull. We call this pull gravity. Gravity is what keeps our feet firmly on the ground and what makes objects fall when we drop them.

Earth's force of gravity also reaches out into the space around it. Earth's gravity holds the Moon in its orbit around our planet.

Earth's Companion

Earth travels through space with a close companion—the Moon. The Moon lies on average about 239,000 miles (384,000 km) away. This seems like a long way, but it's only a small step in space.

The Moon is Earth's only natural satellite. A satellite is a smaller object orbiting around a larger one. The Moon is a different world from Earth. It is drab in colour, has no oceans, no atmosphere, and no life.

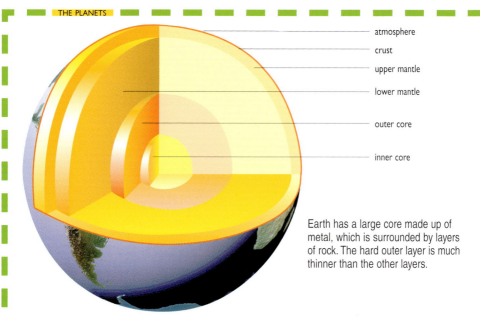

| atmosphere |
| crust |
| upper mantle |
| lower mantle |
| outer core |
| inner core |

Earth has a large core made up of metal, which is surrounded by layers of rock. The hard outer layer is much thinner than the other layers.

EARTH DATA

Diameter at equator: 7,926 miles (12,756 km)

Diameter at poles: 7,900 miles (12,714 km)

Average distance from Sun: 93,000,000 miles (150,000,000 km)

Rotates in: 23 hours 56 minutes

Orbits Sun in: 365¼ days

Moons: 1 (the Moon)

WHAT EARTH IS MADE OF

Scientists who study Earth are called geologists. The word geologist comes from Geos, the Latin word for Earth. From their studies, geologists have found that Earth is made up of several layers.

The centre part of Earth is called the core. The core is extremely hot and is made up of different kinds of metal. The main metals are iron and nickel. In Earth's centre, in the inner core, the metals are solid. But in the outer part of the core, the metals are liquid. No one has gotten near Earth's core, but scientists believe it may get as hot as 11,000° F (6,100° C).

Outside the core is a thick rock layer called the mantle. The rock in the mantle is made up of materials such as magnesium, iron, and silicon. The part of the mantle closest to the core is soft. This soft rock can flow slowly, like hot tar on a road. But the outer part of the mantle is hard and rigid. Although the mantle reaches high temperatures, it is not nearly as hot as Earth's core.

The outermost layer of Earth is made up of a thin layer of rock. This hard rock is called the crust, and it's made of materials such as granite, shale, and marble. We cannot always see the crust because much of it lies under the oceans or is covered by soft ground and plants. The crust is about 25 miles (40 km) thick on land, but only about 6 miles (10 km) thick under the oceans.

Right: Earth is constantly rebuilding itself from the inside through volcanic eruptions. Liquid rock flows from beneath Earth's crust, and solid rock is formed by cooling.

Shocking Methods

Geologists learn about what Earth is like inside by studying waves that move through it during earthquakes. Earthquakes are shakings in underground rock caused by movement inside Earth. These shakings send out ripples called shock waves. The waves travel all through Earth. Scientists in different parts of the world record when the waves reach them and trace the path of the shock waves through Earth. Scientists find that the waves bend at certain depths. The waves bend where they meet a different layer of rock. By studying how the waves bend, scientists can tell where each layer of Earth begins and ends.

site of earthquake

shock waves

inner core

outer core

mantle

Drifting Continents

Earth's surface is not a solid slab of rock. It is made up of many different pieces, like a jigsaw puzzle. All these pieces are moving very slowly, causing continents to drift, oceans to widen, and mountains to form.

If you look at a map of the Atlantic Ocean and the continents on each side, you might notice something. South America looks as if it might fit like a puzzle piece with Africa. And North America looks as if it might fit with Europe in the same way.

In fact, scientists are certain that these continents were joined together a long time ago. But over time, the continents have slowly drifted apart, and the Atlantic Ocean has come between them. This kind of movement is called continental drift, and it happens all over Earth's surface.

There are many examples of evidence of continental drift. For instance, deposits from glaciers that existed hundreds of millions of years ago have been found in warm places such as India, Australia, Africa, and South America. The glacial deposits probably mean that these continents were once in a cold place, possibly close to the South Pole. In the same way, certain fossils found in North America show that our continent was probably once near the equator. Fossils are the remains of living things.

MOVING PLATES

To understand how continental drift takes place, think about the way Earth is made up. The crust and top part of the mantle are solid rock. But the rock layer underneath is soft. Heat from deeper down in the mantle

150 million years ago

These drawings show how the surface of our planet has changed over time.

This map of the world shows the seven large plates and several smaller ones that make up Earth's surface. The red lines show the boundaries between the plates, and the arrows show the directions in which the plates are moving.

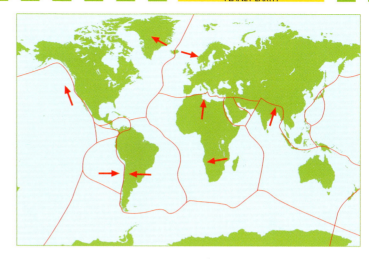

makes the soft rock soft and starts it moving, just like a radiator starts warm air moving in a room. As the soft rock layer moves, it causes slabs of the hard rock above it to move. Scientist call these slabs plates.

The map above shows the main plates that make up Earth's surface. Most of the continents sit on separate plates. The main boundaries between plates are located in the middle of the oceans.

100 million years ago

50 million years ago

present day

THE SPREADING SEAS

The continents are not the only parts of Earth's surface that are moving. The ocean floors move as well. In most oceans, the floor is moving in different directions in different places. For example, the eastern and western parts of the Atlantic Ocean floor sit on different plates that are moving in opposite directions. This means that the ocean floor is spreading apart.

Plates are spreading apart in many parts of the ocean floor. Where this is happening, molten, or liquid, rock pushes its way upward from deep underground. As the molten rock comes up, it spreads east and west. Then it cools and fills in the gap between the existing plates. The cooled rock will in turn be forced to spread apart by new molten rock forcing its way upward. This process is called sea-floor spreading.

Where sea-floor spreading is happening, a mountain range forms underwater. The build-up of new plate material forms underwater ridges along plate bounderies. A mountain range like this runs down the middle of the Atlantic Ocean and is called the Mid-Atlantic Ridge.

SPLITTING CONTINENTS

Sometimes molten rock pushes its way upward through a continent. The molten rock forces the con-

This picture shows the northern part of the Red Sea. The sea is slowly getting wider due to sea-floor spreading.

Below: Movements in hot rock underneath solid plates cause the plates to shift. When plates push against each other in the middle of a land mass (as at left), they push the land in between upward, creating mountains. When molten rock wells up through the ocean floor (as at right), new plate material forms along a mid-ocean ridge.

continental plates colliding, mountains formed

ocean plate

ocean

plate movement

currents in hot rock

tinent to split apart. This process is happening in the Great Rift Valley in Africa. In time, the valley will widen and become flooded with the sea. Then a new ocean will form and grow wider and wider.

WHEN PLATES COLLIDE

New plate material is constantly forming and spreading out in the middle of the oceans. But in other places, plate material is being destroyed. This happens when two plates collide with each other. One plate dips below the other and returns to Earth's interior, melting as it does so.

When an ocean plate collides with a continental plate, the ocean plate dips down. The continental plate rides up over it. This causes the land to wrinkle up and form a mountain range. Rock in the ocean plate melts and may force its way up through the range to form volcanoes. The Andes Mountains on the western edge of South America were formed in this way and have many volcanoes.

The Himalayas are the highest mountain range on Earth. The mountains formed when two plates collided.

Below: When plates collide at the ocean edge, the ocean plate is forced downward. A trench forms under the sea, while the land wrinkles up to form mountains, often with volcanoes.

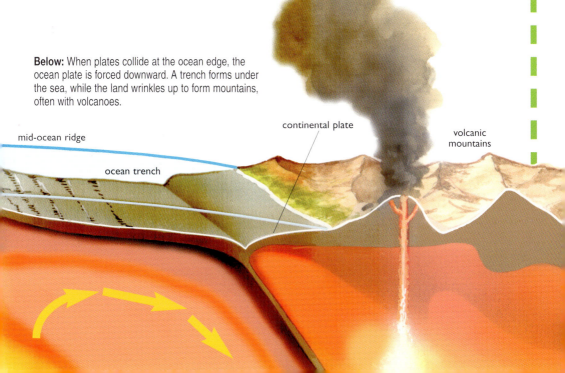

mid-ocean ridge

ocean trench

continental plate

volcanic mountains

Volcanoes and Earthquakes

Movement of the plates that make up Earth's crust can cause volcanic eruptions and earthquakes. Both are among nature's most amazing spectacles, and they can be extremely dangerous.

Molten lava pours over the rocks like a fiery waterfall in Hawaii. It is coming from Kilauea, one of the most active volcanoes on Earth.

When a volcano erupts, great fountains of red-hot liquid rock called lava spurt out through an opening in Earth's crust. Volcanoes also send out great clouds of ash. Lava pours out and runs along the ground in fiery rivers. After lava cools, it turns into hard rock. Most volcanoes erupt many times over a long period of time. Volcanoes gradually grow into mountains, as layers of lava rock build up.

There are hundreds of volcanoes around the edge of the plates beneath the Pacific Ocean. They form what is called the Ring of Fire. Other volcanoes have formed in places such as Hawaii, Iceland, and Africa.

A view from above the excavated remains of Pompeii

Vesuvius and Pompeii

The most famous volcanic eruption in history took place in the year A.D. 79, when Mount Vesuvius erupted in Italy. Its billowing ash cloud buried the nearby town of Pompeii and killed thousands of people. At the time, Pompeii was one of the most magnificent cities in the Roman Empire. After the eruption, the city lay forgotten for nearly 1,700 years. Excavations of the ruins began over 50 years ago, uncovering ancient buildings, streets, and even yhe remains of humans buried by the ash from Mount Vesuvius.

Volcanic eruptions can cause destruction, but the lava they produce also creates new land. Some volcanic eruptions that occur on the ocean floor produce enough lava to eventually form large mounds that rise out of the water and become islands. The Hawaiian Islands began to form in this way about 30 million years ago.

On land, many small and large mountains were once erupting volcanoes. Mount Rainier, the largest mountain in the state of Washington, formed from a volcano. After the volcano became inactive, or stopped erupting, plant and animal life began to inhabit the newly formed land.

This town in Italy has been devastated by a severe earthquake. Nearly every year, earthquakes kill large numbers of people, and destroy many homes around the world.

THE SHAKING EARTH

Like volcanic eruptions, most earthquakes take place along the edges of plates. When plates move over or past one another, they do not move smoothly. Their edges are rough and often lock together. The plates try to move but cannot. Pressure builds up in the rocks until they give way, and the plates move with a sudden jerk. This makes the rocks shake, causing an earthquake.

This satellite picture of the coast of California shows the famous San Andreas Fault. A fault is a fracture in Earth's crust where sections of rock are rubbing against each other. This fault has been responsible for many earthquakes. It runs from the bottom right to the top left of the picture.

An earthquake produces tremors, or waves, that travel through the underground rocks to the surface. When the waves reach the surface, they can cause enormous destruction. They shake houses to pieces, split roads apart, and hurt or kill large numbers of people. In the 1976 earthquake in Tangshan, China, as many as 250,000 people perished.

Earth Rocks

Many kinds of rock make up Earth's crust, both on land and beneath the sea.

Most of Earth's rocks are made of one or more minerals. Minerals are solid substances in the earth that are not part of plants or animals. Each mineral is made up of a separate element or a combination of elements. There are three main kinds of rock on Earth—igneous, sedimentary, and metamorphic. Each kind of rock is formed in a different way.

As certain igneous rocks cool, mineral crystals, such as the tourmaline crystal above, form in the rocks.

COOLED ROCK

Granite is an example of what is called an igneous rock. The word igneous means formed by fire. Granite forms when magma, or hot liquid rock, cools and becomes solid under the ground. It cools slowly, and this allows the minerals in it to grow coarse crystals.

Basalt is another common kind of igneous rock, forms from magma. Basalt is formed when magma has found its way to the surface through a volcano. On the surface, it cools quickly, and the minerals it contains have little time to grow. This is why basalt contains only very tiny crystals.

Basalt and granite are two of the most common rocks found in Earth's crust. They are both igneous rocks.

basalt

Slate is a metamorphic rock. It formed from a sedimentary rock called shale. Then heat and pressure changed the shale into slate.

slate

granite

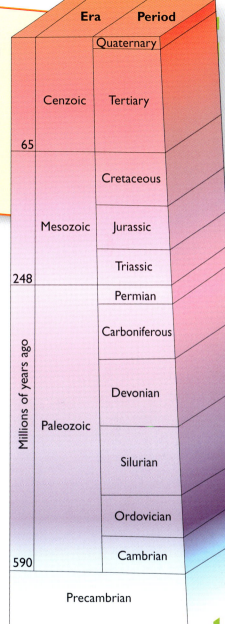

Layers of Time

By studying the different layers in sedimentary rocks and the fossils they contain, geologists can piece together a detailed geological history of Earth. Geologists can tell approximately when the layers were formed, and this gives them the geological time scale. It begins about 600 million years ago, at the time when fossils began appearing in large quantities in the rocks. Geological time is divided into intervals called eras and periods.

Era	Period
	Quaternary
Cenzoic	Tertiary
65	
	Cretaceous
Mesozoic	Jurassic
	Triassic
248	
	Permian
	Carboniferous
Paleozoic	Devonian
	Silurian
	Ordovician
	Cambrian
590	
Precambrian	

Millions of years ago

LAYERED ROCKS

Other kinds of rock are formed from materials such as mud and sand. These materials are washed off the land by rivers and settle in the bottom of ancient seas. The settled material is called sediment. Layers of sediment build up and become pressed together. Over time, the layers turn into solid rock. We call this kind of rock sedimentary rock.

Common sedimentary rocks are shale, which forms from layers of mud, and sandstone, which forms from layers of sand. Some kinds of sedimentary rock formed when ancient seas dried up and left behind the chemicals that were dissolved in them. Limestone is a common example. Chalk is made up of the same chemical. But it is made up of fossils, or the remains of tiny creatures that lived and died in the ancient seas.

CHANGED ROCKS

A third kind of rock is called metamorphic rock. The word metamorphic means having to do with change. Metamorphic rocks are rocks that have been changed by heat and pressure underground. Slate and marble are common metamorphic rocks. Slate was once shale, and marble was once limestone.

The Changing Landscape

The face of Earth is constantly changing due to many forces, including the weather, water, wind, and glaciers. Even the highest mountains will eventually be worn down into little hills.

Stalagmites rise from the floor and stalactites hang from the roof of a limestone cave.

Erosion created the Grand Canyon in Arizona. For millions of years, the Colourado River has been cutting into the rocks so that in places the Canyon is a mile (1.6 km) deep. The layers in the sides of the Canyon tell us that the rocks are sedimentary.

The wearing away of Earth's surface is known as erosion. It is going on all the time. Weather plays a major part in the process, attacking Earth with rain, snow, cold, and heat. Over time, weather breaks down and loosens rock from Earth's surface. This stage of the erosion process is called weathering.

Changes in temperature help cause weathering. During the day, the heat from the Sun makes the outer layers of rocks expand rapidly. At night, they cool down quickly and shrink. This constant expanding and shrinking eventually causes the layers to flake off. Frost has a similar effect. It freezes water that has trickled into tiny holes in the rocks. The water expands as it freezes into ice, and in time it forces off flakes of rock.

ATTACK BY WATER

The landscape is attacked by water in a number of ways. Mountain streams flow swiftly and carry along stones that scrape the rocks in the riverbed. Streams and rivers also gradually dissolve away some of the minerals in the rocks that they flow through.

When river water attacks limestone, it hollows out deep channels and caves in the rock. In these caves, the constant dripping of water creates great stone "icicles" that hang from the roof. Drippings from the icicles create columns on the floor. We call the icicles stalactites and the columns stalagmites.

Humans Changing Earth's Surface

The human race is also changing the face of Earth. Farming, logging, and overgrazing by livestock all cause Earth's crust to erode. For instance, using plows to farm land loosens soil and breaks it down. The soil is then more easily blown or washed away. Logging forest trees also causes erosion. When trees are removed, they no longer hold down the soil and protect it from rain. In the same way, overgrazing by livestock destroys the grass and other plants that hold down soil.

Humans are destroying vast expanses of rain forest every year.

Rivers play another important part in erosion. They help transport material formed by weathering, such as mud and sand. Most often, rivers carry this material, known as sediment, into the sea.

Waves from the sea help push pieces of rock and sand onto sandy beaches. At rocky shorelines, the waves carry particles that crash into rock formations. The force of the waves carves out arches and caves in the rock.

ATTACK BY WIND

In desert regions, wind is a major cause of erosion. It picks up particles of sand and flings them against rocks. This sandblasting gradually wears away the rocks.

ATTACK BY ICE

Glaciers are another cause of erosion. A glacier is a large body of ice and snow that moves very slowly down mountains or over land. Glaciers form in cold places, such as near the North and South Poles, and in high mountains. As it moves across land, a glacier carries broken rock and scrapes away soil in its path.

The Mississippi River dumps about 500 million tons of sediment into the sea every year. The sediment has built up to form an area of land called the Mississippi Delta where the Mississippi empties into the Gulf of Mexico.

The Watery Earth

Most of Earth's surface is covered by the water of the oceans. The largest oceans are the Pacific, the Atlantic, and the Indian Oceans. The Pacific Ocean is twice as big as the Atlantic and covers about a third of Earth's surface.

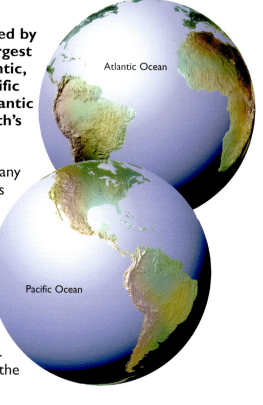

Atlantic Ocean

Pacific Ocean

The water in the oceans has many chemicals dissolved in it and tastes salty. The main chemical in ocean water is sodium chloride, which is the same as ordinary salt we use at home.

Fresh water, which does not contain salt, is found in rivers and lakes on land. It is also found in high mountains and at the North and South Poles in the form of ice. There are traces of fresh water in the air in the form of water vapour.

Water remains permanently frozen as ice near the North Pole.

Waves that have travelled across oceans can grow to a great height when they reach the shore.

THE OCEAN SURFACE

Waves ripple across the ocean surface most of the time. They are created by the wind. In stormy weather, when winds blow strongly, waves can rise up to more than 40 feet (12 m) high.

Winds also cause great currents of water to flow through the oceans. These currents follow the same path year after year. There are warm currents and cold currents. One of the best-known warm currents is the Gulf Stream in the Atlantic Ocean. It travels north along the East Coast of the United States. Westerly winds that blow toward north-western Europe are warmed as they cross the Gulf Stream. These warmed winds help to create mild temperatures in nortwestern Europe.

The level of the water in the oceans does not stay the same. Twice a day, the level rises and falls. This movement is called the tides. Tides are caused by the gravity of the Moon. As the Moon passes over an ocean, it tugs at the water on Earth. This makes the water level rise and causes a high tide. When the Moon moves on, the water level falls to a low tide.

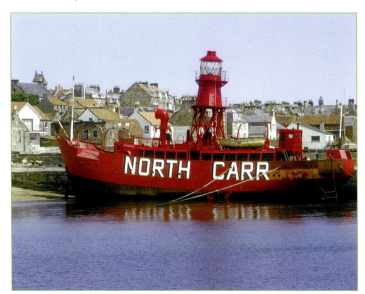

A lightship is stranded ashore at low tide. In a few hours, the tide will flow back in and refloat the ship. Along some coasts, the tide rises and falls as much as 50 feet (15 m). Along others, the tide hardly changes at all.

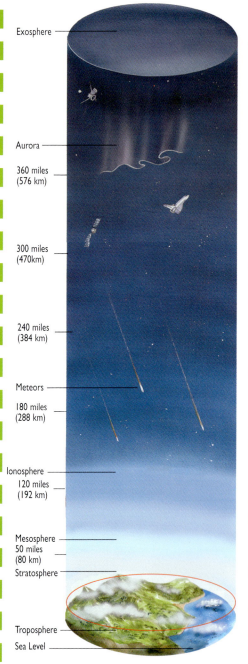

Exosphere

Aurora

360 miles
(576 km)

300 miles
(470km)

240 miles
(384 km)

Meteors

180 miles
(288 km)

Ionosphere
120 miles
(192 km)

Mesosphere
50 miles
(80 km)

Stratosphere

Troposphere

Sea Level

Earth's Atmosphere

The layer of air around and above us makes life possible on Earth. It gives us oxygen to breathe, it helps keep us warm, and it protects us from dangerous rays from outer space.

The layer of gases we call the atmosphere, or air, is a mixture of many gases. Only two gases are present in large amounts. One is nitrogen, which makes up 78 percent of our air. The other is oxygen, which makes up 21 percent. Other gases are present in only very small amounts. They include argon, carbon dioxide, helium, and sulphur dioxide.

Earth's atmosphere stretches to a height of about 400 miles (650 km) above the surface of Earth. The air is thickest near Earth's surface because of the weight of all the air above it. The air becomes thinner higher up. Eventually, the atmosphere fades away into outer space.

The atmosphere is divided into a number of different layers. The bottom layer is called the troposphere. This is where most of our weather happens. The troposphere is about 12 miles (18 km) thick near the equator, but only about 5 miles (8 km) thick near the North and South Poles.

Ninety-nine percent of the air in Earth's atmosphere is found in the bottom two layers, the troposphere and the stratosphere.

THE OUTER LAYERS

The next layer up is the stratosphere, which extends to a height of about 30 miles (48 km). Within the stratosphere is a thin layer of a gas called ozone. The ozone layer is very important to life on Earth because it blocks most of the dangerous ultraviolet rays that come from the Sun. Too much exposure to ultraviolet rays can burn our skin.

The mesosphere is above the stratosphere and extends to about 50 miles (80 km) above Earth. Temperatures are extremely low in this layer. Parts of the mesosphere can be as cold as −100° F (−73° C).

The uppermost layer in Earth's atmosphere is the thermosphere. The air in the thermosphere is extremely thin and grows thinner as it extends toward outer space. The lower part of the thermosphere is called the ionosphere. In the ionosphere, air is present in the form of ions, or electrically charged particles. The displays of coloured light we call the aurora borealis, or Northern and Southern lights, take place in the ionosphere. It is also in this layer that meteors burn up and form fiery streaks of light.

Astronauts on a space shuttle took this photograph of a sunset from space. It shows how dust in Earth's atmosphere turns the sky orange and red during a sunset.

Acid Rain

Sulphur dioxide escapes into the air naturally when volcanoes erupt. But large quantities of this gas are also produced when coal and oil are burned to produce electricity. Sulphur dioxide is a major form of air pollution. It is not only harmful to breathe, it also causes acid rain. In the atmosphere, the gas combines with moisture to form little drops of acid. This falls to Earth as acid rain. Acid rain has caused serious problems in many parts of the world. It kills life in rivers and lakes, kills trees, and attacks the stonework of buildings.

Weather and Climate

The weather on Earth changes from day to day, place to place, and season to season. The main elements that affect our weather are the air's temperature, the amount of moisture in the air, and how the air is moving.

The weather affects all our lives in so many ways that scientists study it closely. The science of the weather is called meteorology, and the people who study it are called meteorologists. They gather information about the weather from weather stations all over the world and from satellites in space. With the help of powerful computers, meteorologists try to predict how the weather will change. Then they issue a forecast that tells us what the weather will probably be like in the next few days.

TEMPERATURE

Three of the most important things meteorologists measure are temperature, pressure, and humidity. The temperature means the degree of heat or cold in the air around us. The temperature depends on how much heat each part of Earth receives from the Sun. Because Earth is round, different places receive more direct sunshine than others, and this makes some places hotter than others.

Because Earth's axis is tilted in space, different parts of Earth are tilted at different angles to the Sun throughout the year. This causes the temperature to change from season to season.

This picture of Earth has been taken by a weather satellite in infrared light to record heat. The continent of Africa stands out because it is very hot.

A warm beam of sunlight reaches Earth most directly at the equator, so this area is always hot. Farther from the equator, sunlight becomes less direct, so temperatures grow colder to the north and south.

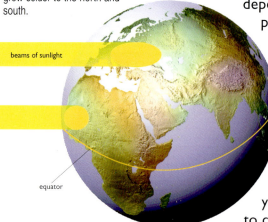

beams of sunlight

equator

PRESSURE

When air becomes hot, it grows lighter and rises. When it becomes cold, it grows heavier and sinks. In the atmosphere, air is rising and sinking all around Earth. This creates air currents, or winds. The movement of large masses of warm and cold air brings about changes in the weather.

Meteorologists keep track of air movements by measuring the pressure of the air, or the force with which it presses down. When warm air rises, it produces low pressure. When cold air sinks, it produces high pressure.

HUMIDITY

Humidity is the amount of water in the air. Water is present in the air in the form of a gas we call water vapour. Water gets into the air when the Sun heats up water in the rivers and oceans and causes it to evapourate, or turn into vapour. Evapouration is part of a never-ending process in which water comes and goes between Earth's surface and the atmosphere. We call this process the water cycle.

In the water cycle, water constantly moves between Earth's surface and the atmosphere.

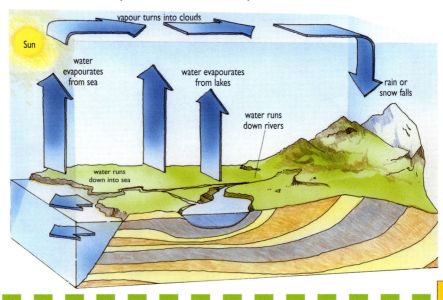

vapour turns into clouds

Sun

water evapourates from sea

water evapourates from lakes

rain or snow falls

water runs down rivers

water runs down into sea

Life on Earth

A fossil trilobite. This species lived on Earth between about 225 and 600 million years ago.

Earth is home to at least 1½ million different species, or kinds, of living things—plants, animals, and simpler life-forms such as bacteria. Life is found practically everywhere on Earth, from the sweltering tropics to the icy poles.

Simple forms of life first appeared on Earth billions of years ago in the early oceans. But it was not until about 600 million years ago that a real "explosion" of life began. This happened in the oceans at the start of the Cambrian period of Earth's history. Life did not appear on land until about 200 million years later. Plants appeared first, followed by insects and amphibians (animals that can live both on land and in water).

As years went by, new species appeared. By about 150 million years ago, reptiles were the main life-form. These included the dinosaurs, or "terrible lizards." Birds also appeared around this time. About 65 million years ago, the dinosaurs and many other species died out. A new class of animals began to flourish. They were the mammals. But it may not have been until about 3 million years ago that early humans appeared on Earth.

Dinosaurs roamed our planet for millions of years.

LIVING KINGDOMS

Biologists, scientists who study living things, class life-forms, or organisms, into groups called kingdoms. The main ones are the plant and animal kingdoms. But some living things are neither plants nor animals. Fungi, such as mushrooms, are not plants because they cannot make their own food. They are placed in a separate kingdom. So are bacteria, which are microscopic organisms, or microorganisms. Other simple organisms, called protozoa (meaning "first animals"), are also placed in a separate kingdom.

PLANT LIFE

Plants are living things that can make their own food. They make food from sunlight, carbon dioxide gas from the air, and water. The process is called photosynthesis.

Like all living things, plants are made up of tiny units called cells. Certain types of algae are the simplest plants. They are very small and are made up of only a single cell. Most plants are made up of many kinds of cells.

Fungi

Dryad's Saddle

Fly Agaric

Morel

Algae

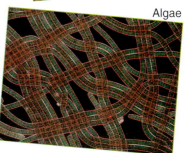

Flowering plants, such as these sunflowers, are complex plants made up of many kinds of cells.

Earth's Moon

Nearly every night, the Moon rises into the sky and helps lighten our darkness. It outshines the stars, the planets, and all the other heavenly bodies in the night sky. At its brightest, it casts shadows. The Moon also looks much bigger than the other heavenly bodies.

The Moon is actually much smaller than Earth—about as big across as the United States is wide. It looks bigger and brighter than the planets and stars because it is so much closer to Earth. It lies only about 239,000 miles (384,000 km) away. The next nearest body to Earth is the planet Venus, which lies more than 26 million miles (41 million km) away.

The Moon is always there in the night sky. This is because it moves with Earth as Earth travels through space. It circles around Earth in a constant path, or orbit. It takes the Moon about a month to circle Earth, and the word month comes from an old form of the word Moon.

The Moon is Earth's only natural satellite. A satellite is a small body that circles around another, larger body. Some planets, such as Jupiter and Saturn, have many satellites, or moons, circling around them.

We know more about the Moon than about any other heavenly body. Astronomers have studied it closely through telescopes for hundreds of years. Space scientists have sent probes to take close-up photographs and land on its surface. And astronauts have travelled to the Moon, walked on its surface.

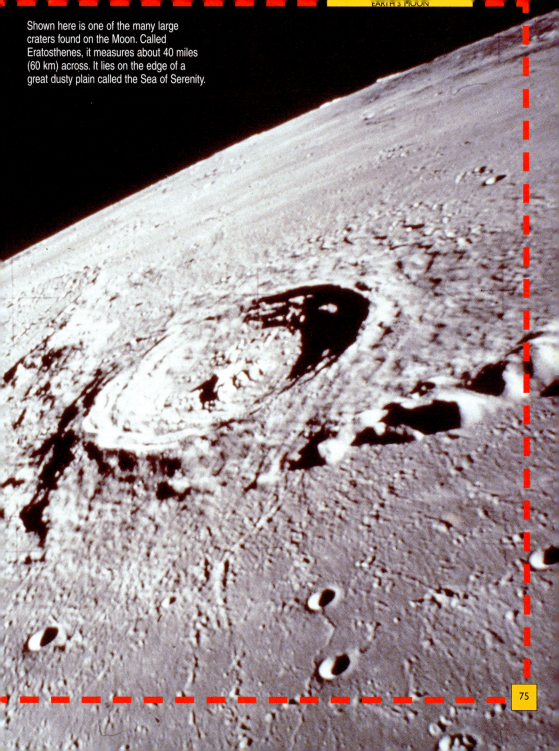

Shown here is one of the many large
craters found on the Moon. Called
Eratosthenes, it measures about 40 miles
(60 km) across. It lies on the edge of a
great dusty plain called the Sea of Serenity.

Looking at the Moon

Every month, the Moon circles around Earth. As it does so, it appears to change shape, going from a thin crescent to a full circle and back again.

But the Moon does not really change shape during the month. The changes we see happen because of the way the Moon shines. It does not give out any light of its own. Instead, it reflects, or sends back, light from the Sun.

The shape of the Moon in the night sky depends on how much of its surface we see lit up by sunlight. This changes night by night as the Moon changes its position in space. We call the Moon's changes in shape its phases.

PHASES OF THE MOON

During the phase we call the new moon, we cannot see the Moon at all. At this time, the Moon is positioned on the same side of Earth as the Sun. The Sun lights up only the Moon's far side. The side of the Moon that faces us remains dark.

Moon 1

Sun

Earth

orbit of Moon around Earth

Moon 2

Man in the Moon

Long before an astronaut became an actual man on the Moon, people talked about the Man in the Moon. This is because, with a little imagination, you can make out a human face in the features visible on the surface during the full moon. Dark areas form the two eyes and mouth of the Moon. Or maybe you can imagine the face of a Woman in the Moon, or the body of a rabbit with long ears.

Over the next few days, the Moon starts to become visible as its position changes. At first we see only a region around the edge of the Moon lit up by the Sun. We call this a crescent moon. Gradually, the crescent grows and grows. A week after the new moon, we see half of the Moon's face lit up. It looks like a semi-circle. We call this phase the first quarter.

A week after the first quarter, we see the whole face of the Moon lit up. This phase is the full moon. The Moon has now travelled halfway around Earth since the new moon. It lies in the opposite direction in the sky from the Sun.

Crescent Moon

Last Quarter

Full Moon

First Quarter

Crescent Moon

In the days after the full moon, we gradually see less and less of the Moon's face lit up. After a week, we see only half of it. It looks like a semi-circle again. This phase is the last quarter. After another week, we can see only a slim crescent. Then the Moon moves back into position between Earth and the Sun and disappears from view. It becomes a new moon once again.

The Moon takes 29½ days to go through its phases from one new moon to the next. When it appears to be growing in size from new to full, we say it is waxing. When it appears to be shrinking in size from full to new, we say it is waning.

An eclipse of the Moon. The Moon has moved into Earth's shadow in space. But you can still see the Moon faintly. This is because some sunlight still reaches it after passing through the Earth's atmosphere.

EARTHSHINE

During the crescent moon, most of the Moon's face is dark. But it is not completely dark. If you look carefully, you can see that it gives off a pale grayish light. This light has been reflected from Earth and is known as earthshine. A popular name for it is "old moon in the new moon's arms."

IN THE SHADOWS

Just as you cast a shadow on the ground, planets and moons cast shadows in space when they block the Sun's light. Earth casts a shadow, and so does the Moon.

Sometimes the Moon passes into the shadow of Earth. We call this an eclipse of the Moon, or a lunar eclipse. Lunar eclipses happen about once or twice a year. The Sun, Earth, and the Moon line up exactly in space, with Earth positioned between the Sun and the Moon.

A lunar eclipse always takes place during a full moon. As the Moon moves slowly into Earth's shadow, we see less and less of it lit up. After about an hour, the Moon is completely in Earth's shadow. But it does not disappear from view. It still gives off a dim pinkish light.

The Moon remains in Earth's shadow for more than

Moon Madness

In ancient times, people held strange beliefs about the Moon. If you gazed at the Moon for too long, they said, you would go crazy. The word lunatic, meaning a crazy person, comes from luna, the Latin word for Moon. Another old belief was that at the time of a full moon, some men could turn into savage, wolflike creatures called werewolves, which would spend the night killing people.

1½ hours. Then it takes about another hour to come out of the shadow completely. From start to finish, an eclipse of the Moon can last up to about 3½ hours.

Sometimes the Sun, the Moon, and Earth line up so that the Moon lies between the Sun and Earth. Then the Moon casts a shadow over part of Earth. We call this an eclipse of the Sun, or a solar eclipse. When this happens, day turns suddenly into night. But because the Moon casts only a small shadow that just reaches Earth, a solar eclipse can be seen only from a small part of Earth. It lasts only for a few minutes. Solar eclipses are not as common as lunar eclipses.

NEAR SIDE AND FAR SIDE

The shape of the Moon in the night sky changes all the time. But the features you see on the Moon always stay the same. The face of one full moon looks exactly the same as the face of the next full moon, and the next, and the next. This is because, from Earth, we only see one side of the Moon—the near side. We never see the other side, the far side.

A total (complete) eclipse of the Sun. Here the Moon has passed in front of the Sun and blocked most of its light. But we can now see the outer atmosphere of the Sun. We call this the corona, which means crown.

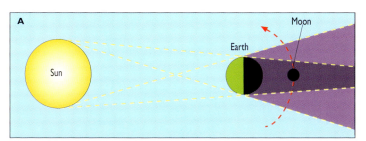

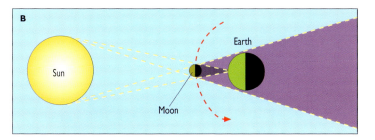

A lunar eclipse (A) occurs when Earth is directly between the Sun and the Moon. A solar eclipse (B) occurs when the Moon is between Earth and the Sun.

Mercury and Venus

Mercury and Venus are two of the planets in our solar system—the family of bodies that circle around the Sun. The solar system contains nine planets altogether, and Mercury and Venus are the only planets closer to the Sun than our home planet, Earth. Being so close to the Sun makes Mercury and Venus much hotter than Earth. In fact, temperatures on the two planets rise so high that they would melt metals such as tin and lead.

Like Earth, Mercury and Venus are made up mostly of rock. We call them terrestrial, or Earth-like, planets. Mars is the other terrestrial planet. But Mercury and Venus are quite different from Earth in most other ways. And they are also quite different from each other. For example, Venus is nearly the same size as Earth, but it is more than twice as big as Mercury. While Venus is surrounded by a thick atmosphere, or layer of gases, Mercury has only a small trace of an atmosphere.

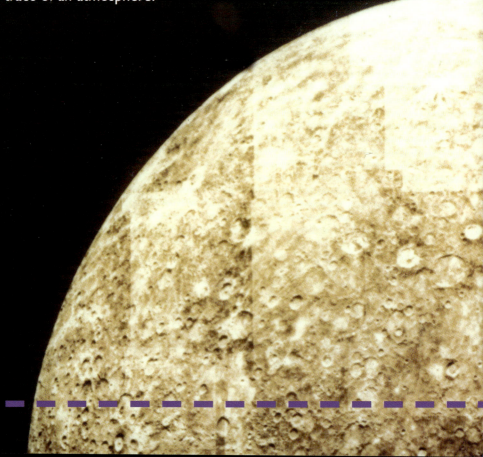

Both Mercury and Venus have been known to astronomers for thousands of years. They can be easily seen with the naked eye, since they often shine brighter than the brightest stars. From our point of view on Earth, both planets stay quite close to the Sun. This means that we can see them only at sunrise or sunset.

Astronomers knew very little about Mercury and Venus until scientists began sending space probes to them. Even the most powerful telescopes on Earth show few features on Mercury's surface because the planet is so small and so far away. Venus is closer and larger, but we cannot see any of its surface from Earth because thick clouds always cover the planet.

Space probes, however, have taught us a great deal about the two planets. We know that Mercury is covered with craters and looks much like the Moon, while Venus is a land of huge volcanoes, unusual formations, and vast plains.

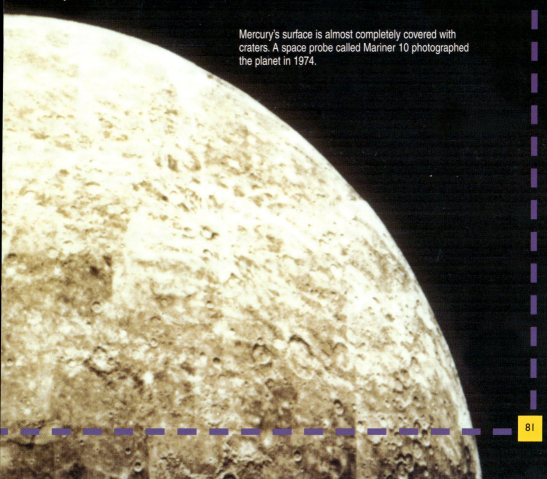

Mercury's surface is almost completely covered with craters. A space probe called Mariner 10 photographed the planet in 1974.

Because Mercury and Venus circle closer to the Sun than Earth does, they always appear in the sky near the Sun. This means we can see them only just before sunrise in the east or just after sunset in the west.

Before sunrise, the sky is becoming lighter, but we can still see Mercury and Venus. They look like bright stars. At this time, we call them morning stars. Just after sunset, the sky is becoming darker, and Mercury and Venus again appear as bright stars. At this time, we call them evening stars.

SHOWING PHASES

As Mercury and Venus circle around the Sun, we see them change in size and shape. Their size appears to change because their distance from Earth is changing all the time. The closer they are to Earth, the bigger they appear in the sky.

The shapes of Mercury and Venus appear to change for another reason. At different times, we see more or less of the planets' surfaces lit up by the Sun. Like all the planets, Mercury and Venus give off no light of their own. They only reflect light from the Sun. Sometimes just a small part of their surface is lit up. Then the planets appear as a thin crescent. At other times, we see half or more of the surface lit up. We call the changing shapes of the planets their phases. They are like the phases of the Moon, but we cannot see

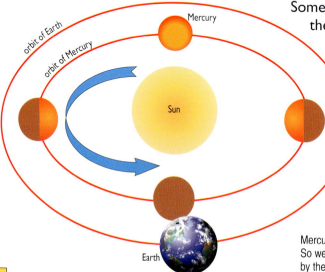

orbit of Earth

orbit of Mercury

Mercury

Sun

Earth

Mercury circles the Sun inside Earth's orbit. So we see different parts of its surface lit up by the Sun at different times.

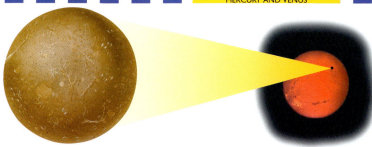

them with the naked eye. We can spot them only in telescopes.

The black dot on the Sun's surface is Mercury. It is making a transit of the Sun. Venus makes transits too, but not as often as Mercury.

PLANETS IN TRANSIT

Once in a while, Mercury or Venus passes directly between Earth and the Sun in space. Then we see the planet pass over the Sun's surface. This is called a transit.

These times when a planet lines up between Earth and the Sun are rare. This is because the planets orbit, or circle, the Sun in different planes. Imagine the Sun on a sheet of paper, with Earth orbiting around it on the paper. Mercury and Venus do not tend to travel on the same imaginary piece of paper—they orbit the Sun in different planes. From Earth, they seem to travel slightly above or below the Sun.

Transits of Mercury happen about 15 times every century. They always occur around early May or mid November. Transits of Venus happen even less often, about twice every century.

Below: Earth orbits the Sun in one plane (top). Mercury and Venus orbit the Sun in a different plane (bottom).

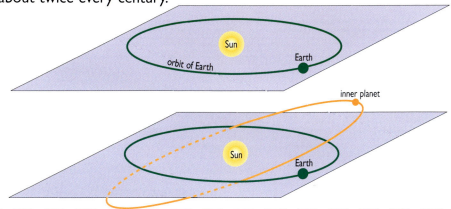

Planet Mercury

Mercury is a planet of many extremes. It is the planet closest to the Sun, the second smallest planet, and the fastest traveller around the Sun. The ancient Romans named it after the speedy messenger of their gods.

The ancient Romans knew Mercury as both an evening star and a morning star. But they thought it was two different bodies. They called it Mercury as an evening star, and Apollo as a morning star.

Mercury is a small planet. With a diameter of 3,031 miles (4,878 km), it is a little over a third the size of Earth.

RACING AROUND THE SUN

Mercury travels in its orbit, or path, around the Sun at an amazing speed of 107,000 miles (172,000 km) per hour. It takes the same amount of time as 88 days on Earth to orbit the Sun once. This is only one-fourth of the time Earth takes to travel around the Sun.

Mercury travels in an oval-shaped, or elliptical, orbit. At times it gets as close to the Sun as 28 million miles (46 million km). At other times, it wanders nearly more than 43 million miles (60 million km) away.

Mercury is the nearest planet to the Sun, at an average distance of about 36 million miles (58 million km). It takes 88 Earth-days to circle the Sun once. Its immediate neighbour in space is Venus.

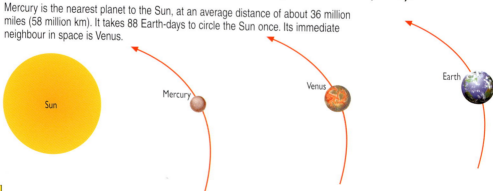

Ancient symbol for the
planet Mercury

Mercury's barren surface is
covered by thousands of craters.

In a Spin

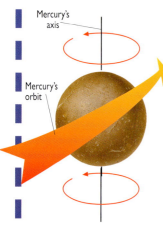

Mercury's axis

Mercury's orbit

Mercury spins around slowly on its axis, taking nearly two Earth-months to rotate once.

Like all the planets, Mercury moves in space in another way. It rotates, or spins around on its axis like a top.

A planet's axis is an imaginary line that runs through its centre from its north pole to its south pole. Earth rotates on its axis once every 24 hours, or 1 day. Mercury spins around relatively slowly, taking nearly 59 Earth-days to rotate once.

This slow spin means that each "day" and "night" on Mercury are very long. Imagine that you travel to Mercury and arrive at a place where the Sun is just rising. You will see the Sun climb very slowly into the sky. About 44 Earth-days will pass before it is "noon" on Mercury, with the Sun high overhead. And another 44 days will pass before the Sun sets. At this time, the

Because of Mercury's slow rotation and its fast orbit, a day on Mercury—the time between sunrise and sunset—lasts 88 Earth-days.

planet will have travelled once around the Sun. Mercury's long day is then followed by an equally long night. The Sun does not rise again for another 88 days.

HOT AND COLD

Because of its long days and nights, Mercury is both a very hot and a very cold place. During the long day, the Sun beats down on part of the planet for nearly three Earth-months at a time. It bakes the planet's surface to 840° F (450° C), which is twice as hot as most home ovens.

After the Sun sets, Mercury's surface quickly cools down. Unlike Venus, Mercury has almost no atmosphere to act as a blanket to keep in the heat. All the heat escapes into space, and the temperature falls to about −290° F (−180° C). This is much, much colder than it has ever been on Earth.

STARPOINT

If you lived on Mercury, you would see the Sun appear to change size during the day, as the planet moved closer or farther away from it.

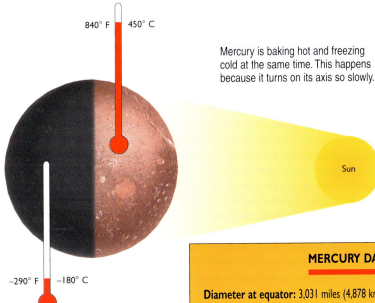

840° F   450° C

Mercury is baking hot and freezing cold at the same time. This happens because it turns on its axis so slowly.

Sun

−290° F   −180° C

MERCURY DATA

Diameter at equator: 3,031 miles (4,878 km)
Average distance from Sun: 36,000,000 miles (58,000,000 km)
Rotates on axis in: 58.7 Earth-days
Orbits Sun in: 88 Earth-days   **Moons:** None

Inside Mercury

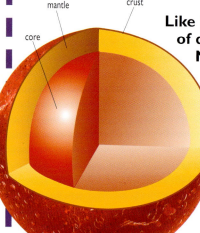

core

mantle

crust

Like Earth, Mercury is a rocky planet made up of different layers. Like Earth's Moon, Mercury is covered by thousands of craters.

Mercury has three main layers. Its outer layer, called the crust, is made up of hard rock. Beneath the crust is a thicker layer of rock known as the mantle. Beneath the mantle is a great ball of iron, which forms the planet's core. Mercury has a large amount of iron for its size. In fact, it has a greater percentage of iron than any other body in our solar system.

Surrounding the planet is a very thin atmosphere. It is only one-trillionth as thick as Earth's atmosphere. A few traces of gas are found near Mercury's surface. The gases present include helium, hydrogen, and sodium.

MAGNETIC MERCURY

Mercury's huge iron core is probably the reason for its magnetism. Magnetism is the force that draws certain metals to iron, and magnetic forces are created when masses of iron rotate. Earth also has an iron core, and the magnetism created is the force that makes a compass always point north. Earth's magnetism is stronger than Mercury's because our planet rotates much faster.

MERCURY'S SURFACE

Craters cover most of Mercury's surface. They were created by lumps of rock from outer space that have crashed down on the planet for billions of years.

Astronomers believe that most of the craters on Mercury were formed about 4 billion years ago, around

Mercury is made up of three main parts—the crust, the mantle, and the core. The planet is not much bigger than the Moon, which is shown below for comparison.

600 million years after the solar system formed. At the time, the solar system was full of rocky lumps, which rained down on planets and moons. Some were meteorites, or lumps of rock that range in size from tiny pebbles to huge chunks thousands of feet across. Others were larger lumps that measured miles across. These are known as the asteroids, which still circle the Sun in a large band between Mars and Jupiter.

When a lump of rock strikes a planet's surface, it creates a crater. The largest meteorites and asteroids made craters hundreds of miles across. These craters resemble the Moon's large craters, with raised rims and deep floors. They also have a small range of mountains in the middle.

When meteorites hit Mercury, masses of rocky fragments were thrown out. Large fragments fell back to the surface and made their own smaller craters. Particles of dust settled in spoke-like patterns around the craters. They reflect sunlight well and show up as shining crater rays.

STARPOINT

Mercury's faint atmosphere is less than one trillionth as thick as Earth's atmosphere.

A close-up view of Mercury's surface. All the craters were made by lumps of rock from outer space that rained down on the planet billions of years ago.

This map of Mercury is based on pictures taken by the Mariner 10 space probe. Craters have been named after great composers, writers, and artists.

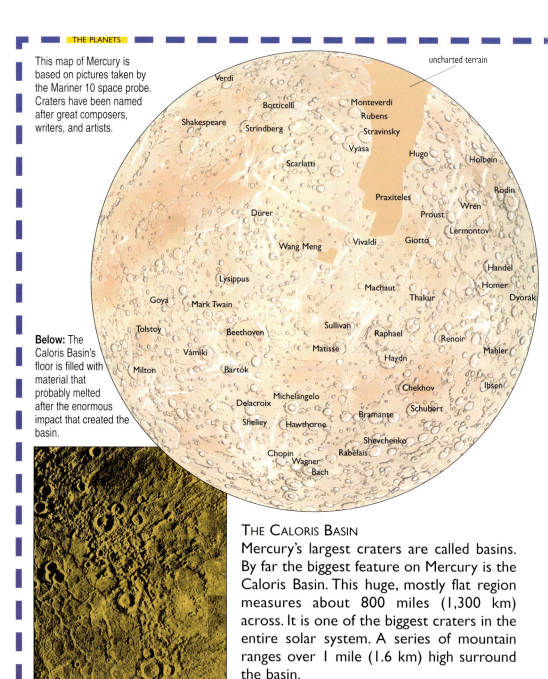

uncharted terrain

Verdi
Botticelli
Monteverdi
Rubens
Shakespeare
Strindberg
Stravinsky
Vyasa
Hugo
Holbein
Scarlatti
Rodin
Praxiteles
Wren
Dürer
Proust
Lermontov
Wang Meng
Vivaldi
Giotto
Lysippus
Handel
Machaut
Homer
Goya
Mark Twain
Thakur
Dvorák
Tolstoy
Sullivan
Raphael
Beethoven
Renoir
Vámiki
Matisse
Mahler
Milton
Haydn
Bartók
Chekhov
Ibsen
Michelangelo
Delacroix
Schubert
Shelley
Hawthorne
Bramante
Shevchenko
Chopin
Rabelais
Wagner
Bach

Below: The Caloris Basin's floor is filled with material that probably melted after the enormous impact that created the basin.

THE CALORIS BASIN

Mercury's largest craters are called basins. By far the biggest feature on Mercury is the Caloris Basin. This huge, mostly flat region measures about 800 miles (1,300 km) across. It is one of the biggest craters in the entire solar system. A series of mountain ranges over 1 mile (1.6 km) high surround the basin.

The Caloris Basin was formed when an asteroid about 60 miles (100 km) across slammed into Mercury billions of years ago. The force of the impact made the planet's surface ripple, like water in a pond ripples when a stone is thrown into it. The ripples became the rings of mountains around the basin.

The impact of the asteroid also sent shock waves through the underground rocks. The waves travelled all the way to the other side of the planet, then shook and cracked the surface there. The result is a strange, bumpy landscape unlike any other part of Mercury. Astronomers call it the "weird terrain."

PLAINS AND CLIFFS

Not all of Mercury is heavily cratered. About 60 percent of the planet's surface is covered by large flat regions, or plains. Also known as planitia, they are smooth like the flat plains on the Moon we call maria, or seas. They probably formed when lava, or hot liquid rock from underground, forced its way up through cracks in the surface. Then the pools of lava cooled and hardened to form the smooth plains.

In several places on Mercury there are steep cliffs known as scarps. Scarps are long, rounded cliffs that rise from 1,000 feet (300 m) to nearly 2 miles (3.2 km) high. Scarps stretch from 10 to 300 miles across Mercury's surface. No other planet has these formations, and neither does our Moon. One scarp called Discovery Rupes stretches for more than 300 miles (500 km) near the southern edge of the planet. Astronomers think that scarps are blocks of Mercury's crust that were forced upward when the planet cooled and shrunk after it formed.

What's in a Name?

Astronomers have named the various features on Mercury's surface after famous people and ships. Craters have been named after great composers (Beethoven and Vivaldi), artists (Michelangelo and Renoir), and writers (Tolstoy and Shelley).

Scarps were named after famous ships of exploration. Discovery Rupes was named after the ship that Captain James Cook sailed from Great Britain to the Hawaiian Islands in 1778.

STARPOINT

Beethoven is the largest crater on Mercury that we know of. It measures about 400 miles (650 km) across.

Planet Venus

Venus is the planet that travels nearest to Earth. It shines brighter than any other object in the sky, except for the Sun and the Moon. The ancient Romans named Venus after their goddess of love and beauty.

Most people think of Venus as the evening star. We often see it shining brilliantly in the western sky just as the Sun goes down. We do not usually see it as a morning star because it appears so early in the morning, before sunrise.

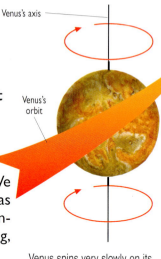

Venus's axis

Venus's orbit

Venus spins very slowly on its axis, and in the opposite direction from the other planets.

VENUS IN MOTION

Venus travels in an almost perfect circle as it orbits the Sun at a distance of about 67 million miles (108 million km). Its path is only slightly elliptical. It takes about 225 Earth-days to travel once around the Sun.

Like all the planets, Venus also rotates on its axis. But it spins very slowly, taking 243 Earth-days to rotate once. So Venus it takes longer to rotate once than it does to travel around the Sun. Venus's day (243 Earth-days) is longer than its year (225 Earth-days).

All the other planets rotate toward the east. On Earth, this means that the Sun appears to rise in the east and set in the west. However, Venus spins in the opposite direction, toward the west. So if you lived on Venus, you would see the Sun slowly rise in the west and set in the east.

Venus is only a little smaller than Earth. Its diameter is about 400 miles (650 km) less than Earth's.

A LAYERED PLANET

All we see of Venus from Earth is its cloudy atmosphere. But beneath the atmosphere is a rocky planet with a makeup similar to that of Earth. Like Earth and Mercury, Venus has different layers. Venus's atmosphere is so thick that it can be considered one of the planet's layers.

Underneath the atmosphere is the hard rocky outer layer called the crust. Scientists don't know its exact thickness, but it is probably 10–30 miles (15-50 km) deep. In the deepest part of the crust, the rock is probably so hot that it is molten, or liquid.

Beneath the crust is a thick layer of heavier rock called the mantle. At Venus's centre is a large metal core, made up mainly of iron and nickel. It may be partly molten, like Earth's metal core.

On Earth, the metal core produces Earth's magnetism. This happens because our planet rotates rapidly in space. Venus, however, rotates too slowly to create magnetism.

The thick clouds that cover Venus look white in ordinary photographs, but coloured patterns show up in images sent back by space probes.

Right: Beneath its layer of clouds, Venus has a hard, rocky crust. Underneath is another rocky layer called the mantle, and beneath this a metal core.

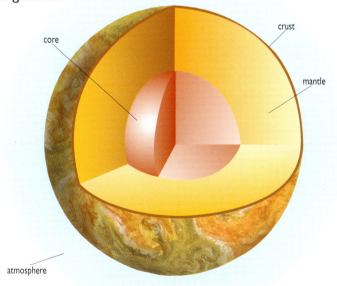

core

crust

mantle

atmosphere

STARPOINT

Venus is the hottest planet in our solar system. Temperatures at the surface rise as high as 900° F (480° C).

Venus's Atmosphere

Venus's thick atmosphere has a crushing pressure. The clouds that cover the planet are made up of tiny drops of acid.

The atmosphere of Venus is quite different from the atmosphere we have on Earth. Earth's atmosphere is made up mainly of nitrogen and oxygen, the gas we must breathe to stay alive. Venus's atmosphere contains only a little nitrogen and no oxygen. The main gas is carbon dioxide, a very heavy gas. All that carbon dioxide makes the whole atmosphere heavy. Gravity causes it to press down on Venus's surface with more than 90 times the force of Earth's atmosphere. In other words, the atmospheric pressure on Venus is 90 times what it is on Earth.

Up in the Clouds

On Earth, the highest clouds form about 6 miles (10 km) above our planet's surface. Clouds in Venus's atmosphere extend much higher. Above Venus, the main cloud layers form at about 30 miles (50 km) above the planet's surface. The clouds extend up to about 65 miles (100 km) above the planet's surface.

Earth clouds are made up of tiny droplets of water or ice. On Venus, clouds are made up mainly of tiny drops of sulphuric acid. Sulphur is released into the atmosphere when volcanoes erupt on the surface. Chemical reactions between sulphur gas and water vapour (water in its gas form) change the sulphur into acid.

Several layers of cloud make up Venus's atmosphere. They occur at different levels and circulate in different directions.

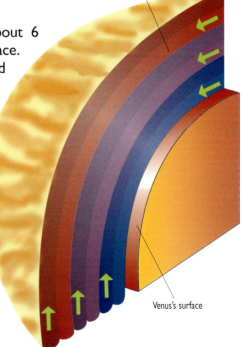

cloud layers

Venus's surface

THE RUNAWAY GREENHOUSE

Besides providing oxygen for us to breathe, Earth's atmosphere also helps keep our planet warm. It acts like a greenhouse by letting in heat from the Sun and keeping some of it from escaping back into space. Many scientists believe that Earth's greenhouse effect is causing our average temperatures to rise. They think this is happening because larger amounts of heavy gases, such as carbon dioxide, are building up in our atmosphere and trapping more of the Sun's heat.

Venus's atmosphere also has a greenhouse effect. In fact, it has turned the planet into a world hotter than an oven. On Venus, the large amounts of carbon dioxide in the atmosphere trap most of the Sun's heat. This keeps the planet at temperatures as high as 900° F (480° C). The temperature does not vary much from place to place or from day to night because the heavy atmosphere acts as such a good blanket, or greenhouse. Some scientists believe that by learning more about the greenhouse effect on Venus we can better understand the same process on our own planet.

Right: The clouds on Venus reflect most of the Sun's light. Only a small amount gets through. But this heat gets trapped by the carbon dioxide in Venus's atmosphere.

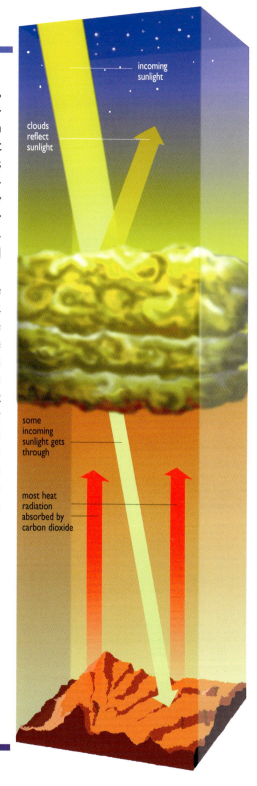

incoming sunlight

clouds reflect sunlight

some incoming sunlight gets through

most heat radiation absorbed by carbon dioxide

VENUS DATA

Diameter at equator: 7,521 miles (12,104 km)
Average distance from Sun: 67,200,000 miles (108,200,000 km)
Rotates on axis in: 243 Earth-days
Orbits Sun in: 224.7 Earth-days   **Moons:** None

Venus's Surface

Under its thick cloud covering, Venus has some amazing landscapes. There are a few highland areas, but most of the planet consists of vast rolling plains dotted with volcanoes.

The ancient symbol for the planet Venus You can see it a few days after the new moon.

Volcanic mountains, rolling hills, and ancient lava flows are the main features of Venus's surface. This picture was produced by computer from radar images sent back by the Magellan space probe.

Venus may be similar to Earth in size, but its surface is quite different. More than two-thirds of Earth's surface is covered by the water of the oceans. Land areas, or continents, cover less than one-third of Earth's surface.

Venus has no great oceans or any water at all on the surface. The planet is too hot. If there ever were oceans on Venus, they would have boiled away a long time ago.

There are two large highland areas on Venus that we can think of as continents, like the continents on Earth. The largest one, Aphrodite Terra, lies close to Venus's equator. This continent is about the same size as South America. It features some spectacular volcanoes. One

volcano, called Maat Mons, rises about 5.5 miles (9 km) high, about as tall as Earth's Mount Everest.

ISHTAR TERRA

Ishtar Terra, the other main continent on Venus, lies farther north. Two main features dominate Ishtar Terra, which is about the size of Australia. One is a mountain range known as Maxwell Montes. Some of the peaks in this range soar as high as 7.5 miles (12 km). Ishtar Terra's other main feature is Lakshmi Planum—a vast plateau, or high plain, surrounded by mountains.

Venus has several other smaller highland areas. Two of them are known as Alpha Regio and Beta Regio. Some highland areas are topped by ancient volcanoes.

This map of Venus was prepared from pictures taken by space probes that used radar to look through the planet's cloud covering.

THE ROLLING PLAINS

Highland regions cover less than one-fifth of Venus's surface. The rest is covered by relatively flat plains. Most of these plains consist of gently rippling hills—we call them rolling plains. There are also lower and flatter plains regions, generally known as lowlands. Guinevere Planitia is one of the largest lowland regions, stretching for about 4,000 miles (7,000 km).

CRATERS ON VENUS

Like all the planets, Venus has been bombarded by lumps of rock called meteorites for billions of years. When meteorites struck the planet, they dug out craters in its surface. Venus has hundreds of craters, but nowhere near as many as can be found on Mercury.

The main reason for Venus's smaller number of craters is that it has a young surface. Great volcanic eruptions have changed the landscape in the past few hundred million years. In the time scale of our solar system, this is fairly recently. These "recent" lava flows covered up all the old craters. So the craters we see on Venus were created within the last few hundred million years. In contrast, most of the craters on Mercury are billions of years old.

There are few small craters on Venus's surface. Small meteorites burn up in Venus's thick atmosphere before they reach the planet's surface. So it is mostly larger meteorites that crash into Venus's surface.

Meteorites blast out masses of rocky surface material when they strike a planet. On Venus, some material settled around the crater, and some was blown away by wind. In places, we can see streaks formed by the wind-blown dust. Rippling dunes, like the sand dunes in deserts on Earth, are also visible.

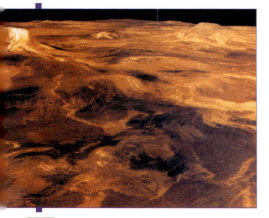

This view of Venus's surface was taken by the Magellan probe.

Shaping the Surface

Volcanoes have been the main force at work in shaping Venus's surface. Hundreds of them dot the landscape, and scientists believe many are still active.

On Earth, we know of three main kinds of volcanoes—cinder cones, shield volcanoes, and composite volcanoes. Cinder cones give off thick, sticky lava that does not flow very far. These volcanoes usually form a steep, cone-shaped mountain. Mount St. Helens in Washington State is a cinder cone.

Shield volcanoes give off very runny lava, which flows a long way. These volcanoes grow into a broad, flat mountain. The volcanoes of Hawaii are shield volcanoes. Composite volcanoes are a combination of the other two kinds. Mount Fuji in Japan is a composite volcano.

Venus's biggest volcanoes measure up to 300 miles (500 km) across and rise several miles high. But most are just a few miles across and a few hundred feet high.

Most of the volcanoes on Venus are shield volcanoes. Over time, the volcanoes poured out runny lava over and over again. This lava flowed widely. When it cooled and hardened, it formed the great rolling plains of Venus's surface.

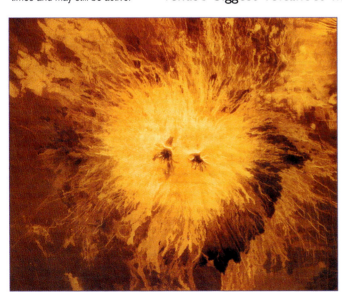

This is one of the many volcanoes found on Venus. It is at least 1 mile (1.6 km) high and measures more than 250 miles (400 km) across. It has erupted many times and may still be active.

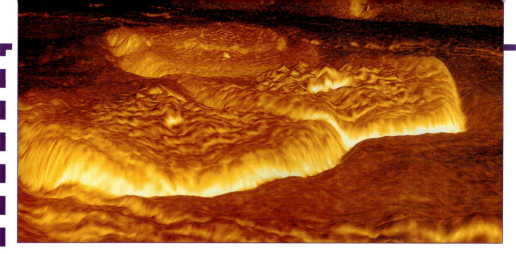

Above: This is a close-up picture of a pancake dome, a feature not seen anywhere else in the solar system. It measures about 15 miles (25 km) across and is about 2,500 feet (800 m) high.

Below: This corona on Venus was produced by volcanic action. It measures about 250 miles (400 km) across.

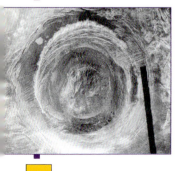

CHANNELS AND PANCAKES

The volcanoes of Venus have created many interesting features on the planet's surface. Flowing lava formed snaking channels that are hundreds of miles long. They look like dried-up riverbeds on Earth.

Eruptions have also created another interesting feature on Venus—the pancake domes. These flat, circular structures are usually tens of miles across and only a few thousand feet high. Nothing like them has been found anywhere else in the solar system.

CROWNS AND SPIDERS

Other features unique to Venus are the tall, circular coronae, which means crowns. Coronae are surrounded by a ring of ridges and troughs. Scientists believe coronae were formed by an upward movement of hot material from deep inside Venus. As the material cooled, the upper layers rose and fell, creating cracks in the surface.

Smaller than coronae but similar to them are Venus's arachnoids. These circular structures look much like spiderwebs, and the word arachnoid means like a spider. Arachnoids measure from 30 to 138 miles (50 to 230 km) across. They have a volcanic peak in the centre, surrounded by a network of fine cracks. Arachnoids may have been created by molten lava pushing up from below.

THE TORTURED CRUST

The fine cracks, or fractures, that cover Venus's coronae and arachnoids were formed when the rocks in the crust moved and split. More widespread movements in Venus's crust have created other surface features, including mountain ranges. Movements of rock up or down along faults, or weaknesses, in the crust have created long troughs and ridges. Some of these troughs and ridges run for thousands of miles.

Even more prominent are the valleys, or rifts, that follow fault lines in many places. Wide rift valleys are found both in the plains and in the highland regions. The continent Aphrodite Terra is riddled with rift valleys. One called Diana Chasma is over 250 miles (400 km) wide in places.

Strange features called tesserae were also caused by movements in Venus's crust. These criss-cross patterns of ridges and grooves appear on no other body in the solar system.

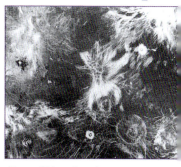

Above: This region of Venus's surface is about 1,000 miles (1,600 km) wide. It shows a number of the spidery features called arachnoids.

Probing Mercury and Venus

Astronomers knew very little about what Mercury and Venus were like until they sent space probes to them.

All the close-up pictures we have of Mercury were taken by the NASA space probe Mariner 10, which visited the planet in 1974.

Mariner 10 set out for Mercury in November 1973, launched from Cape Canaveral, Florida. But it did not fly to Mercury by the most direct route. First it was sent to fly by Venus. It did this to pick up speed. It used the pull of Venus's gravity to make it move faster.

It took nearly three months for Mariner 10 to reach Venus, where it took over 4,000 pictures of the cloud-covered planet in February 1974. Venus's gravity speeded it up and boosted it into a trajectory (path) that took it to Mercury in March. It flew as close as 435 miles (700 km), taking pictures that showed how much Mercury looked like the Moon.

After leaving Mercury, Mariner 10 looped around the Sun before flying past Mercury once more in September, though much farther away (more than 25,000 miles, or 40,000 km). Six months later, it returned again, this time skimming only about 200 miles (300 km) above the planet's surface.

The space probe Mariner 10 had twin television cameras that took pictures of Mercury and Venus. Picture signals were transmitted by the dish antenna. Twin solar panels, each nearly 9 feet (2.7 m) long, supplied power to the craft. A sun shade protected the main body of the spacecraft from the Sun's heat.

American scientists launched the first successful probe to Venus in 1962. Mariner 2 sent back the first reports of Venus's very high temperature. Russian scientists achieved the next success when they parachuted instruments from their probe Venera 4 into Venus's atmosphere in 1967. It confirmed that the planet was very hot and reported that it had a thick atmosphere of mostly carbon dioxide.

Above: This Russian Venera probe explored Venus in the 1970s. The ball-shaped capsule at the bottom of the craft was dropped into the atmosphere and floated down to the surface by parachute.

RUSSIAN LANDINGS

Three years later, the Russian probe Venera 7 actually landed on Venus's surface. It sent back information about Venus's atmosphere, but no pictures. In 1975, however, Venera 9 and Venera 10 landed and sent back pictures. The pictures showed a number of what looked like volcanic rocks.

By this time, the American probe Mariner 10 had photographed Venus from space on its way to Mercury in 1974. At its closest, the probe swooped to within 4,000 miles (6,400 km) of the planet's surface. Its photographs showed bands of clouds swirling in the thick atmosphere.

Mariner 10 took this picture of Venus in ultraviolet light. In this kind of light, the bands of clouds swirling about in the planet's atmosphere show up clearly.

RADAR IMAGES

While space scientists were sending probes to Venus, astronomers began using other ways to look at the planet. Astronomers used a system called radar (radio detecting and ranging) to send radio waves to Venus. The radio waves were reflected back and then collected and used to create radar images of the planet. These images showed general features of the planet's surface. Highland areas and fault lines were visible.

SCANNING FROM ORBIT

The next leap forward in our knowledge about Venus came in December 1978, when Pioneer Venus 1 began orbiting Venus. It sent back radar images of the planet, produced a map of its surface, and measured temperatures in its upper atmosphere. It discovered the main features of the landscape, such as the two main continents and the rolling plains.

The Pioneer Venus 1 probe was the first to use radar to look at Venus's surface. Launched in May 1978, it went into orbit around Venus seven months later.

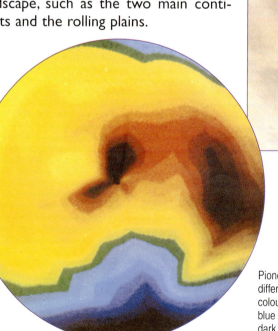

Pioneer Venus 1 took this picture, which shows differences in temperature on Venus. The bright colours show the hotter side lit by the Sun. The blue regions show lower temperatures on the dark side of the planet.

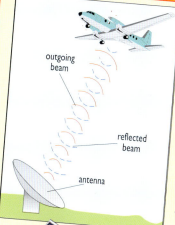

outgoing beam

reflected beam

antenna

Below: The Magellan probe orbited Venus for more than four years. It weighed over 3½ tons and had a dish antenna 12 feet (3.7 m) across. Its twin solar panels measured about 11 feet (3.5 m) across.

MARVELOUS MAGELLAN

The Pioneer Venus probe showed us what a fascinating place Venus is. But its instruments could not make out much detail. So space scientists decided to send a more advanced probe to the planet. Magellan was launched in May 1989. After 15 months, it reached its target, then it went into orbit around Venus in August 1990. Radar images from Magellan provided scientists with details about Venus's surface features. Magellan operated successfully for about four years, before falling into Venus's atmosphere in October 1994 and burning up like a shooting star.

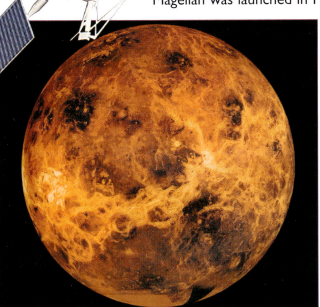

Left: The Magellan probe showed Venus to be unlike any other body in the solar system.

Mars

Mars looks like a bright reddish-orange light in our night sky. The planet's unique colour reminded ancient peoples of fire and blood. The Romans named the planet after Mars, their god of war. Because of its distinct colour, Mars is often called the Red Planet.

Mars is the fourth planet in the solar system, going out from the Sun. That makes it one of our closest neighbours. Mars, along with Mercury, Venus, and Earth, is one of the four terrestrial planets. Terrestrial means Earth-like. The terrestrial planets are all made up of rocky layers surrounding a metal core. Mars also has an atmosphere, or layer of gases, surrounding it. Beneath the atmosphere, ice caps rise at the freezing north and south poles.

Mars's surface has many Earth-like features, including volcanoes, deep valleys, and what appear to be dried-out riverbeds.

Many people once thought that life, even intelligent life, might exist on Mars. But information and pictures from space probes have not shown that life exists on the Red Planet. Conditions on the surface are so harsh that life could not survive. There is no liquid water, the temperatures are too low, and the atmosphere does not contain enough of the gases needed to support life.

However, long ago, Mars was probably much warmer, had a thicker atmosphere, and had flowing water. Astronomers (scientists who study outer space) continue to search for evidence that some form of life did exist on Mars many years ago. Perhaps, as we learn more about our mysterious neighbour, we might find proof that life once existed on the Red Planet.

A close-up view of the Red Planet, Mars

Mars Basics

Mars is one of the smallest planets in the solar system and one of the closest to Earth. It is easy to spot when it shines brightly in our night sky.

As the fourth planet in the solar system, Mars lies in between Earth and Jupiter. Mars's distance from the Sun varies, from about 128 to 155 million miles (206–250 million km). At that distance, it takes longer to orbit, or circle, the Sun than Earth does. Mars completes one orbit in 687 Earth-days—nearly two Earth-years.

CLOSE ENCOUNTERS

Mars has a diameter, or distance across, of about 4,220 miles (6,800 km). That's just over half the diameter of Earth. Only Mercury and Pluto are smaller. Despite its small size, Mars is an easy planet to spot in our night sky because of its reddish colour.

Above: Mars is small compared with our own planet. It could fit inside Earth nearly seven times.

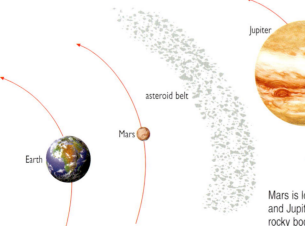

Jupiter

asteroid belt

Mars

Earth

Mars is located in the solar system between Earth and Jupiter. Between Mars and Jupiter is the ring of rocky bodies we call the asteroids.

But Mars is not easily visible all the time. It shines most brightly when Mars and Earth move close together as they travel in their orbits around the Sun. The two planets come closest to each other about every 26 months. Sometimes Mars and Earth come within 35 million miles (56 million km) of each other during these close encounters. Only Venus travels closer to Earth.

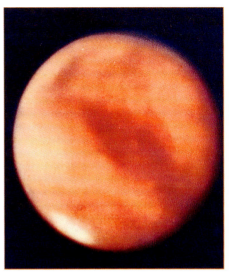

Through a telescope on Earth, dark markings and an ice cap (bottom left) can be seen on Mars.

THROUGH A TELESCOPE

If you looked at Mars through a telescope, you would notice dark and bright areas on its surface. The bright parts are covered by dust, sand, and rocks. They make up about two-thirds of the planet's surface. The dark areas cover about one-third of the planet. Mars's dark parts are often called maria, or seas, although they do not contain water. From Earth, they appear to change in size and position over time. These changes could be caused by sand that blows across Mars and covers up some of the dark areas.

Telescopes also reveal another prominent feature on Mars—white caps at its north and south poles. Like the caps at Earth's poles, Mars's polar caps are made up of frozen ice. The size and appearance of Mars's ice caps changes throughout the year.

MARS'S MAKEUP

Like Earth, Mars is made up of several rocky layers. Its thick crust, or hard outer layer, is about 125 miles (220 km) deep. The crust covers a deeper layer of heavier rock called the mantle, which is about 1,360 miles (2,190 km) thick. At Mars's centre is a core made up mainly of iron. Unlike Earth's liquid metal core, Mars's core is probably solid.

MARS DATA

Diameter at equator:
4,220 miles (6,794 km)

Average distance from Sun:
142,000,000 miles
(228,000,000 km)

Rotates in: 24 hours, 37 minutes

Orbits Sun in: 687 days

Moons: 2

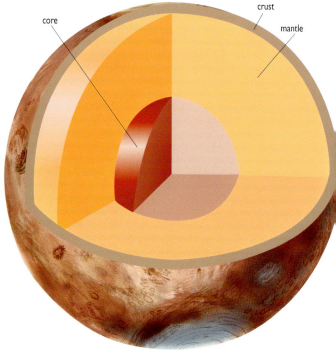

core

crust

mantle

polar cap

DAYS ON MARS

Like all the planets, Mars spins around, or rotates, on its axis as it travels in its orbit around the Sun. A planet's axis is an imaginary line running through it from its north pole to its south pole. Mars rotates once in about 24 hours and 37 minutes in Earth time. One complete rotation equals one day. This means that one day on Mars is just slightly longer than the 24-hour day we have on Earth. A Martian day is known as a sol. Astronomers have divided the sol into 24 Martian hours, so that each Martian hour lasts 61 Earth-minutes and 33 Earth-seconds.

SEASONS ON MARS

As on Earth, the weather on Mars changes regularly over periods of time called seasons. The tilt of Mars's

STARPOINT

Mars takes nearly twice as long to orbit the Sun as Earth does, each of its four seasons lasts nearly twice as long as our seasons.

axis causes the planet to have seasons. Mars's axis tilts at an angle of 24 degrees, just a little more than Earth's axis. This tilt means that different parts of Mars lean toward or away from the Sun as the planet travels in its orbit. The part of the planet tilting toward the Sun receives more of the Sun's heat and grows warmer. When one half of the planet is tilted toward the Sun, the other half is tilted away. So when it is summer on Mars's northern half, it is winter on its southern half.

Above: On Mars, as on Earth, when one half of the planet is in daylight, the other half experiences darkness.

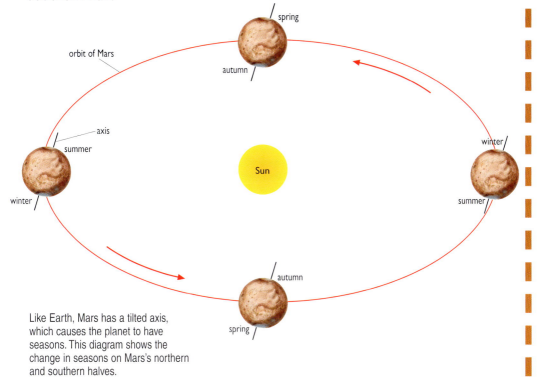

spring

orbit of Mars

autumn

axis

summer

winter

Sun

winter

summer

autumn

spring

Like Earth, Mars has a tilted axis, which causes the planet to have seasons. This diagram shows the change in seasons on Mars's northern and southern halves.

THE MARTIAN MOONS

The American astronomer Asaph Hall discovered Mars's two moons in 1877. He named them Phobos and Deimos after the sons of the Roman god Mars. Both moons are small and have an irregular shape, sort of like a potato. Phobos, the larger moon, measures only about 17 miles (28 km) across at its widest point. Deimos is no more than 10 miles (16 km) across. Craters, or pits in the surface, cover both moons, and Phobos is marked with long grooves. Dusty soil and rocky boulders rest on each moon's surface.

Phobos orbits very close to Mars. In fact, it orbits closer to its planet than any other moon does. It travels at a distance of only about 3,700 miles (6,000 km)

Mars's two small moons, Phobos (above) and Deimos (right). Both have an irregular shape.

STARPOINT

In the 1960s, a Russian scientist named Iosif Shklovskii believed that Phobos was hollow. He thought it might be a huge space station built by intelligent Martians.

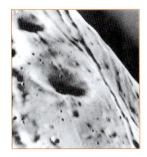

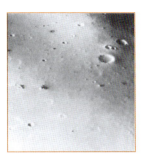

The surface of Phobos (top) is marked with long grooves and deep craters. Deimos's surface (bottom) is also covered with craters.

from Mars. Because it is so close to Mars, Phobos speeds around the planet in less than eight hours, travelling from west to east. In comparison, Deimos takes over a day to orbit Mars. It orbits at a distance of about 12,000 miles (20,000 km) from the planet.

Astronomers think that Mars's potato-shaped moons may have once orbited the Sun as asteroids. Most of these rocky bodies travel around the Sun in what we call the asteroid belt, located between Mars and Jupiter. Phobos and Deimos would have been captured by Mars's gravity when they came close to the planet billions of years ago. Gravity is the attraction, or pull, one body has on the objects around it.

Phobos in orbit around Mars

This picture of Mars shows the icy south polar region. The large circular patch to the right is a great cloud of mist.

Atmosphere and Weather

The weather on Mars is always changing. There are swirling clouds, morning mists, blowing winds, and raging dust storms.

Highs and Lows

While temperatures vary widely across Mars, overall it is much colder than our home planet. Like Earth, Mars is warmest at the equator—an imaginary line around a planet, midway between its north and south poles. And also like Earth, Mars is coldest at its poles. Temperatures on Mars vary from place to place, changing with the seasons. In the Martian summer, temperatures at the equator may rise to nearly 70°F (22°C). But elsewhere, daytime temperatures are generally below freezing. During the winter, temperatures may drop to less than –220°F (–140°C) at the poles. The average temperature on Mars is about –67°F (–55°C).

Above: Dust storms on Mars often begin on flat stretches of land like this one, called Solis Planium.

Compared to Earth, Mars has a very thin atmosphere. It is made up of about 95 percent carbon dioxide and small amounts of other gases, including nitrogen and argon. Mars's atmosphere contains just a faint trace of oxygen—the gas that humans and animals on Earth must breathe to live.

Mars's atmosphere also contains small amounts of water vapour (water in its gas form). As on Earth, the water vapour forms clouds. Clouds on Mars swirl in the atmosphere and cluster around the higher slopes of mountains. Low clouds, or mist, also form in the valleys early in Mars's morning.

MARTIAN WINDS

Winds on Mars blow from the cold parts of the planet to the warmer parts. In general, Martian winds are light, but at times they can blow at speeds of up to 250 mph (400 km/h). These powerful winds whip up fine dust from Mars's surface to form dust storms in the atmosphere. Dust storms on Mars can last for weeks and cover the whole planet, with dust rising up to 15 miles (24 km) in the atmosphere.

Right: Winds on Mars stir up dust on its surface, leading to dust storms in the planet's thin atmosphere.

Mars's Landscape

High volcanoes, deep basins, long channels, and frozen polar caps are just some of the amazing features that cover Mars's surface.

Mars looks very different to the north and south of its equator. Ancient craters, or large pits in the surface, cover most of Mars's southern half. Astronomers believe that many of these craters are billions of years old. The surface of Mars's northern half is probably much younger. Unlike Mars's ancient cratered land, the landscape north of the equator is marked by plains, or flat stretches of land, that formed as recently as 1 billion years ago.

The Hubble Space Telescope orbits Earth at 380 miles (610 km) above our planet. It sends back pictures like this one of the Martian landscape. It observes the planet on a regular basis, following the changes that take place season by season.

THE MARTIAN VOLCANOES

Mars may be one of the smallest planets, but some of the largest volcanoes in the solar system rise on its northern half. Over millions of years, Mars's volcanoes erupted many times, pouring out floods of lava, or hot liquid rock. As the lava spread over Mars's surface, it cooled and hardened to form the planet's northern plains.

THE BIG FOUR

Mars's four largest volcanoes are not shaped like cones with steep sides, like many volcanoes on Earth. Instead, these volcanoes have a wide base and gently sloping sides. Similar volcanoes on Earth are called shield volcanoes. Three of Mars's largest volcanoes rise on a great bulge called Tharsis, near the planet's equator. The three volcanoes, named Ascraeus Mons, Pavonis Mons, and Arsia Mons, are strung out in a line. They are each

STARPOINT

Olympus Mons is 100 times the size of Earth's largest volcano, Mauna Loa, which is located in the Hawaiian Islands.

about 9 miles (15 km) high and at least 250 miles (400 km) across at the base.

As big as they are, these three volcanoes are dwarfed by another volcano to the west called Olympus Mons. At its base, it measures nearly 400 miles (640 km) across. If Olympus Mons were on Earth, it would cover all of the state of Washington and nearly half of Oregon. The giant volcano rises to about 17 miles (27 km) above the surrounding landscape. That's three times the height of Mount Everest, the highest mountain on Earth.

VOLCANOES EVERYWHERE

To the north of the Tharsis Bulge sits another remarkable volcano called Alba Patera. It stands only a few miles high, but it spreads over an area as much as 400 miles (700 km) across. Also in the north is a grouping of three smaller volcanoes, in a region called Elysium.

A number of volcanoes lie near the huge crater called Hellas, on Mars's southern half. The volcanoes near this crater probably erupted when the crater formed as long as 4 billion years ago. In comparison, the four huge northern volcanoes may have erupted as recently as 1 billion years ago.

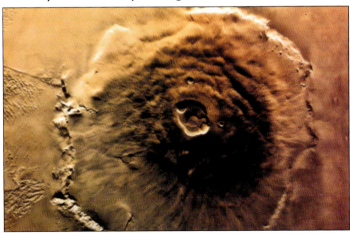

The giant volcano Olympus Mons (left) erupted time and time again, pouring out rivers of lava that flowed over Mars's surface. A model (below) shows what the volcano's crater would look like close up.

STARPOINT

The floor of Hellas is over 3 miles (5 km) lower than the surrounding surface, making it the lowest region on Mars.

This image shows mainly the northern half of Mars. To the left are the three large volcanoes on the Tharsis Bulge. To their right is the great canyon system Valles Marineris.

CRATERS AND BASINS

Mars's southern half looks a lot like our Moon. Large and small craters cover the landscape. Astronomers believe that many of these craters formed about 4 billion years ago, when meteorites hit Mars's surface. Meteorites are rocky lumps from outer space that crash to the surface of another body, such as a planet or moon.

Blowing winds have eroded, or worn away, some of Mars's craters. Other craters have been flooded by lava from ancient volcanic eruptions. The craters vary in size from small pits to large ring structures more than 1,000 miles (1,600 km) across. The largest crater on Mars is the Hellas Basin, measuring as much as 1,200 miles (2,000 km) across.

A 3D view of part of Valles Marineris

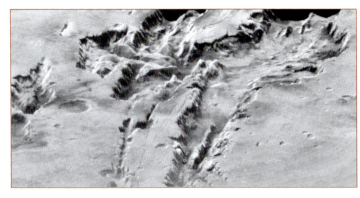

MARS'S GRAND CANYON

An enormous canyon near Mars's equator stretches one fourth of the way around the planet. It is often called the Grand Canyon of Mars, after Earth's famous canyon in Arizona. Its actual name is Valles Marineris. The canyon begins near the great Tharsis bulge and runs east along the equator. Many branches fan out to the north and south of the main canyon.

With a length of about 2,500 miles (4,000 km), Valles Marineris is more than 10 times as long as the Grand Canyon on Earth. It is up to 120 miles (200 km) wide and as much as 4 miles (7 km) deep. In comparison, the Grand Canyon is at most only about 20 miles (30 km) wide and a little over 1 mile (1.6 km) deep.

The Difference

Valles Marineris and the Grand Canyon look similar, but they formed in very different ways. Earth's Grand Canyon was created by the Colourado River. Over hundreds of thousands of years, the river has slowly carved the Grand Canyon out of the rocks it flows over. In contrast, Valles Marineris was created over a much shorter period of time. It probably formed when a part of Mars's crust shifted and caused the surface to crack open.

The Grand Canyon, in Arizona

STARPOINT

Scientists have estimated that early in Mars's history, its surface may have been covered by 1,500 feet (465 m) of water.

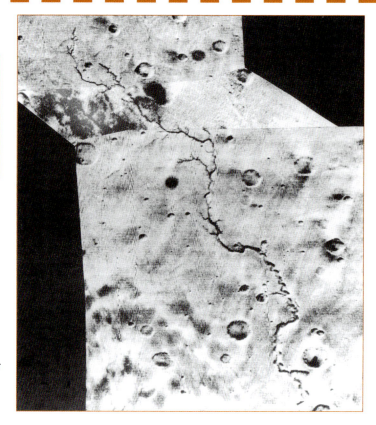

A channel zigzags across the surface of Mars for a distance of about 360 miles (575 km). It looks very similar to the channels that rivers make when they flow on Earth.

CHANNELS

A river did not carve out Valles Marineris, but there are signs elsewhere on Mars that water might have flooded the planet in the past. Winding channels run across different parts of Mars's surface. They appear to be places where rivers may once have flowed.

Some channels on Mars begin in high ground and gradually widen as they wind their way to lower ground, just as rivers on Earth widen as they flow downward. Other channels are not as wide and have smaller channels that branch off. In some channels, surface material has formed into patterns around obstacles such as craters. Astronomers believe that flowing water may have caused these patterns.

Water on Mars?

The Mars we know is a dry planet. But the existence of so many channels leads astronomers to believe that water may once have flooded the surface. Water might have spewed from erupting volcanoes billions of years ago, when Mars's inner layers were probably very hot. These volcanic eruptions would have also released huge amounts of different gases. Together, rising water vapour and other gases would have formed a thick atmosphere. A thicker atmosphere would have kept in more of the Sun's heat, so at one time the whole planet might have been much warmer.

Gradually the gases in Mars's thicker atmosphere would have drifted off into space. Most of the water vapour would have drifted off too, leaving only enough vapour in the planet's thin atmosphere to form clouds and freeze as ice caps at the poles. No liquid water remains on the planet, but water exists as ice in Mars's polar caps and may be frozen underneath the planet's dusty surface.

A deep valley cuts through the surface near Mars's equator and connects with the Amazon Plain (top). It is possible that this flat region was once the floor of a large ocean.

THE POLAR CAPS

For a long time, astronomers have noticed that the ice caps on Mars's north and south poles change in size. When they discovered that Mars has regular seasons, they realized that the polar caps grow and shrink as temperatures change. Both polar caps are biggest during winter, when more ice freezes, and smallest during summer, when more ice melts. Seasons in the north and south of the planet are opposite. This means that when one ice cap is at its largest, the other is at its smallest.

In winter at both poles, gas from Mars's atmosphere freezes on the ground, making the ice caps grow. These winter ice caps are probably made up of a mixture of frozen carbon dioxide and frozen water. When temperatures rise again in the summer, some of the ice turns into gas and goes back into the atmosphere. But it never gets warm enough at either pole to melt the ice caps entirely. The ice cap that remains at the north pole in summer is made up mainly of water ice, like the polar caps on Earth. The summer ice cap at the south pole is probably a mixture of water ice and frozen carbon dioxide.

Mars's northern polar cap has a noticeable spiral pattern, probably caused by swirling winds.

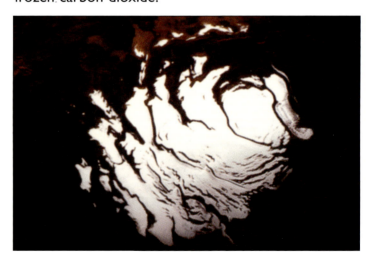

Right: This is what remains of the southern ice cap of Mars during summer. The cap measures about 200 miles (320 km) across. It is built up of layers and cut with deep, curving valleys.

Mars Rocks

Rocks of all shapes and sizes are found on Mars's great plains, amid drifts and dunes of fine red soil.

Before we knew much about Mars's surface, astronomers expected the planet's flat northern plains to be like sandy deserts on Earth. Instead, astronomers discovered that thousands of rocks cover Mars's dusty plains. Martian rocks come in all shapes and sizes. Some are as small as pebbles, while others are as large as boulders. They can be rounded or sharp, dark or bright, pitted or smooth.

A Variety of Rocks

Where do all the rocks that cover the Martian soil come from? Some may have erupted from ancient volcanoes, which would have hurled rocks high into the sky while they poured out hot lava. Other rocks may have broken off from Mars's surface when meteorites struck the planet. Scientists also believe that flowing water on Mars may have deposited rocks on the surface. An example of this kind of rock lines the mouth of what is probably an ancient riverbed on Mars.

Large and small rocks cover much of Mars's rust-red surface, as seen on this plain called Chryse.

The Martian Soil

Dry, dusty soil covers every-thing on Mars's surface. Some of the soil is very fine, possibly as fine as talcum powder. It is blown by the wind into drifts or dunes that are similar to those in deserts on Earth. The Martian soil is rich in iron oxide, the substance that forms rust. The presence of iron oxide in the soil gives Mars its reddish colour.

TESTING THE ROCKS

In 1996, NASA launched a probe called Pathfinder. It landed on Mars's surface in July 1997. Several different kinds of rock surrounded the area where Pathfinder landed. Most of the rocks appeared to have come from volcanic eruptions. Tests on the rocks showed that they were like two kinds of volcanic rock on Earth called basalt and andesite. Pathfinder craft also tested other rocks that looked like a rock called conglomerate found on Earth. This kind of rock forms in riverbeds.

STARPOINT

The Pathfinder mission tested rocks on Mars by bombarding them with radiation. The way the rocks reflected the radiation helped scientists learn about their makeup.

Three different kinds of rocks are marked in this picture, taken by the Pathfinder probe. Red arrows point to rocks that are large and rounded and have a surface worn by the action of the weather. Blue arrows show small, gray angular rocks, while white arrows point to flat light-coloured rocks.

Life on Mars?

People once believed that intelligent beings lived on Mars. We have since learned that Martians do not exist. But many scientists hope to find proof that some form of life did exist on Mars in the past.

In 1877, an Italian observer named Giovanni Schiaparelli studied Mars closely and reported seeing dark lines on the planet. He called them canals. Some people began to believe that intelligent beings had made these canals, or artificial waterways.

In 1894 the astronomer Percival Lowell set up an observatory at Flagstaff, Arizona, to study Mars. Like Schiaparelli, Lowell believed that he could see networks of canals on the Red Planet. Lowell wrote about the possibility of desperate Martians living on a planet that was slowly growing colder and drier. He believed that Martians had built canals to carry precious water from the melting ice caps to warmer areas near the equator. They needed the water, Lowell believed, to grow crops. But they were fighting a losing battle and were slowly dying out.

Other astronomers with equally good telescopes did not find any canals on Mars. They did not support the idea that intelligent Martians lived on the planet. But even in the early 1960s, many people thought that there might be some form of life on Mars.

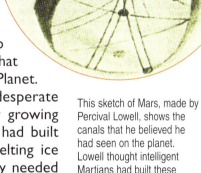

This sketch of Mars, made by Percival Lowell, shows the canals that he believed he had seen on the planet. Lowell thought intelligent Martians had built these canals.

UNINVITED VISITORS
The English writer H. G. Wells took up this theme in his novel The War of the Worlds (1898). He wrote about a dying Martian race that could see how green and fertile Earth was. The Martians decided to send an army

Above: This is a scene from the film of H. G. Wells's *The War of the Worlds*, which was first screened in 1953. It shows the tendril-like limb of a Martian reaching out for a terrified human.

to Earth in terrifying war machines so that they could invade and conquer their neighbouring planet. This book gave more people the idea that intelligent beings might be living on Mars.

In 1938, the American actor Orson Welles produced a radio broadcast of a play based on The War of the Worlds. In the play, Martians were invading New Jersey. Thousands of American radio listeners did not realize they were hearing a play, and they panicked. Some people fled to their churches to pray, others locked themselves in their homes. Hundreds of terrified citizens called their local police. Frightened listeners eventually learned that Martians had not invaded Earth.

CLOSE-UP VIEWS

When space probes began visiting Mars in the late 1960s, astronomers learned that conditions on the planet are too harsh for life as we know it. The probes found no crops, no canals, and certainly no signs of intelligent Martians. However, if Mars was once warmer and had flowing water, life may have existed on the planet in the distant past. It is possible that we may one day find fossils, or the remains of living things, on Mars.

Right: The probes that landed on Mars did not discover canals or intelligent Martians on the planet. Instead, they have shown that conditions on Mars are too harsh to support life.

Missions to Mars

Since 1965, several space probes have visited the Red Planet, sending back pictures of giant volcanoes, misty valleys, wind-blown deserts, and wide craters.

Above: This is what Mariner 4 would have looked like as it approached Mars in July, 1965.

The Soviet Union tried to launch the first space probe to Mars in 1960. Over the next few years, they attempted to launch more probes, but none of them reached the planet. NASA also had problems with its first probe to Mars, Mariner 3, which was launched in November 1964. and never made it to Mars.

Also launched in November 1964, NASA's Mariner 4 became the first space probe to reach Mars. It flew past the planet on July 14, 1965, at a distance of less than 6,000 miles (10,000 km). Mariner 4 sent back information about Mars's atmosphere and took pictures that showed craters on the planet's surface. These craters looked much like the craters on the Moon.

Two more probes, Mariner 6 and Mariner 7, reached Mars in July and August of 1969, coming within 2,200 miles (3,500 km) of the planet. Together, Mariner 6 and Mariner 7 sent back nearly 200 pictures. They showed that Mars had desert-like

Below: Mariner 6 revealed the surface of Mars clearly for the first time. It showed that parts of the planet look much like the Moon.

plains, and many craters. They also photographed the southern polar cap and recorded temperatures as low as −190°F (−125°C).

Mariner 9 took this picture of the mouth of Mars's largest volcano, Olympus Mons.

MARINER 9 IN ORBIT

The next probe to reach Mars, Mariner 9, went into orbit around the planet in November 1971. At the time, Mars was in the grip of a dust storm, which prevented Mariner 9 from revealing much of the planet's surface.

Within days of the Martian skies clearing, Mariner 9 spotted two of Mars's outstanding natural features—its remarkable Valles Marineris and the massive ancient volcano, Olympus Mons. By the time Mariner 9 ran out of fuel, in October 1972, it had sent back more than 7,000 pictures and photographed almost the entire planet. During this time, several Soviet spacecraft had also begun to successfully explore Mars.

The Viking Invasion

Mariner 9 paved the way for an exciting mission called Viking. On this mission, two identical probes were sent to Mars. NASA planned for the probes to release landers—spacecraft that would drop down to Mars and explore its surface.

Viking 1 sped into space in late August 1975, with Viking 2 following in early September. For more than 10 months, the Viking spacecraft cruised through space. Viking 1 went into orbit around Mars in June 1976, followed by Viking 2 a few weeks later. In their orbits, the Viking probes came as close as 950 miles (1,500 km) to Mars's surface to photograph possible places for the landers to drop.

The Viking landers descend to Mars's surface. The landing capsule separates from the Viking orbiter (1). The capsule maneuvers so that its heat shield faces forward (2). A parachute slows it down (3) before the lander separates and uses rockets to brake (4), so that it lands gently (5). The orbiter stays in orbit (6).

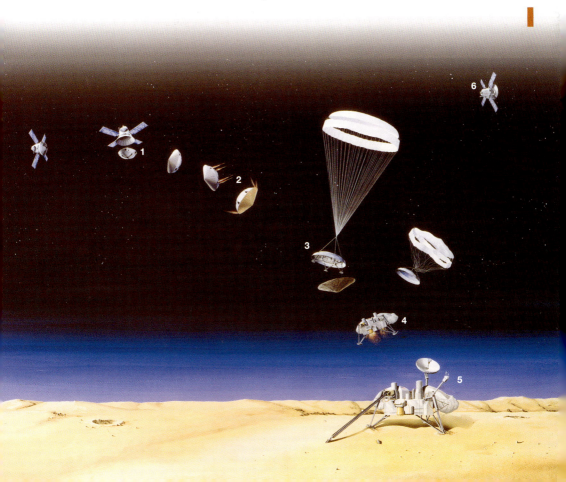

This photograph taken by a Viking orbiter shows the red soil on Mars's surface and white clouds in its thin atmosphere.

TOUCHDOWN

The Viking orbiters released their landing craft to chosen sites in July and September. The Viking 1 lander touched down on a flat plains region known as Chryse, and the Viking 2 lander touched down hundreds of miles away on another part of Mars called Utopia. Surprisingly, the landers' cameras showed that the landscape of both areas looked almost the same.

The landers took pictures, reported on the Martian weather, and dug into the soil. A scoop on each lander's digging arm placed soil samples into a miniature laboratory, where experiments were carried out on them. The experiments tested the soil for traces of life. But no traces were found.

Meanwhile, the Viking orbiters were photographing the planet, sending back the most detailed views of Mars to date. Together, they took more than 51,000 pictures. The final transmissions from the Viking mission were sent in November 1982, when scientists lost contact with the spacecraft.

Right: The Viking orbiter measured about 32 feet (9.8 meters) across the paddle-like solar panels. It had two cameras, which were like the TV cameras used in broadcasting. It took pictures through a series of coloured filters so that colour pictures could be built up. Other instruments measured temperatures and the amount of water vapour in the atmosphere.

Above: As the rover Sojourner explored the Martian surface, it transmitted its findings to Pathfinder, which relayed them back to Earth.

Left: Pathfinder and Sojourner on Mars. In the background are the mountains known as the Twin Peaks.

THE MARTIAN ROVER

NASA did not send another probe to Mars for nearly 15 years. In July 1997, Pathfinder landed on Mars's surface. After Pathfinder landed, it unfolded a set of petal-like solar panels. The panels collected sunlight and used it to power its instruments. Pathfinder's first photographs showed distant mountains, the rim of a crater, and a reddish surface scattered with rocks. The landscape looked very similar to the areas that the Viking probes had observed.

But Pathfinder carried more than cameras. It carried a wheeled rover called Sojourner, which travelled slowly over the Martian surface. This small vehicle inspected nearby rocks and analyzed their makeup. Pathfinder and Sojourner sent back data for nearly three months.

By this time, another NASA spacecraft called Mars Global Surveyor was orbiting Mars. It sent back highly detailed pictures of the surface from space. Then in late 1998 and early 1999 the Climate Orbiter and the Polar Lander launched from Earth. Scientists continue to learn more and more about our mysterious neighbour. Someday, perhaps we will learn whether life has ever existed on the Red Planet.

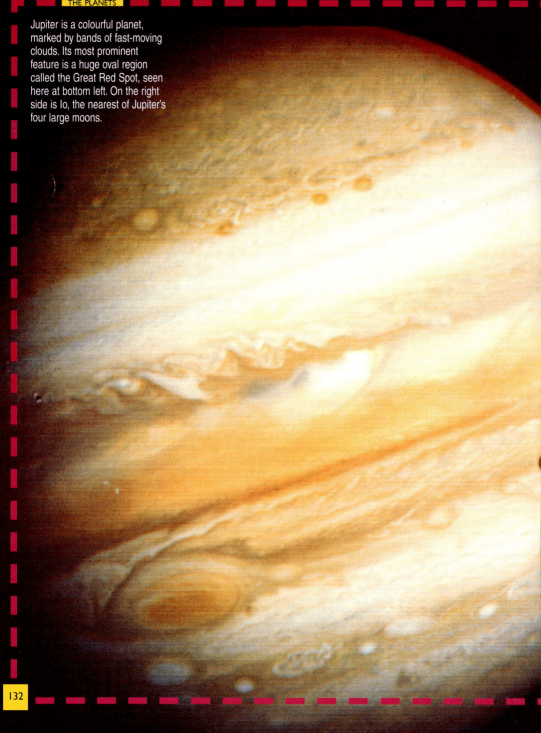

Jupiter is a colourful planet, marked by bands of fast-moving clouds. Its most prominent feature is a huge oval region called the Great Red Spot, seen here at bottom left. On the right side is Io, the nearest of Jupiter's four large moons.

Jupiter

Jupiter is the largest of the nine planets in the solar system—the family of bodies that circle in space around the Sun. Jupiter is so big that all of the other planets could fit inside it easily. It has more than 300 times the mass, or matter, of our own planet, Earth.

Jupiter is quite a different kind of planet from Earth. Earth is one of the four rocky planets of the inner solar system—Mercury, Venus, Earth, and Mars. Jupiter is one of the gas giants of the outer solar system—Jupiter, Saturn, Uranus, and Neptune. The ninth planet, Pluto, is a small ball of rock and ice.

Jupiter is one of the brightest objects in our night sky, after the Moon and the planet Venus. Like all the planets, Jupiter shines because it reflects light from the Sun. In powerful telescopes, we see that Jupiter is a colourful planet, with light and dark bands and white and coloured spots. These are features of Jupiter's atmosphere—the thick layer of gases that covers the planet.

Jupiter is the centre of its own kind of "solar system." At least 16 moons circle around Jupiter, in much the same way the planets circle around the Sun. Jupiter keeps its large family of moons in place with its enormous gravity. Gravity is the attraction, or pull, that a heavenly body has on objects on or near it. Jupiter's moons have an amazing variety of features.

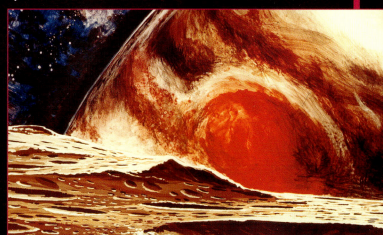

Right: An artist's impression of the Great Red Spot, seen from Amalthea, a tiny moon that orbits much closer to Jupiter than Io.

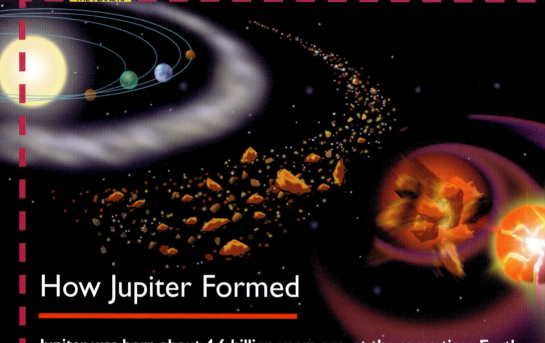

How Jupiter Formed

Jupiter was born about 4.6 billion years ago, at the same time Earth and the other planets formed. It became a different kind of planet from Earth because it formed so much farther away from the Sun.

The solar system formed out of a huge cloud of dust and gas, mainly hydrogen and helium. Over time, the cloud shrank into a huge spinning ball, with a disk of gas and dust circling around it.

The ball of gas at the centre gradually became smaller as it collapsed under the pull of gravity between its particles. As the ball collapsed, it heated up. It eventually started to glow, and in time it began to shine as a star, the Sun.

As the new Sun shone, it blasted surrounding layers of gas away from it in the form of a furious solar wind. This process continued for millions of years.

PLANETS FORM

Meanwhile, changes were taking place in the disk spinning around the Sun. The inner part of the disk was quite warm. There, lumps of rock and metal were forming out of smaller bits as they kept bumping into one another. In time, these lumps became bigger and bigger until they formed into the rocky Earth-like planets of the inner solar system.

Farther out, the disk was much colder. The bits of matter found there were lumps of ice and frozen gases. Over time, these lumps also grew into large bodies—the five planets of the outer solar system.

All the while, the Sun had been blowing gases away from the inner part of the disk into the colder outer part. These gases, mainly hydrogen and helium, formed a great cloud around the icy bodies that had formed there. These bodies began to attract more and more gases, growing bigger in the process. The body we call Jupiter happened to be in the thickest part of the gas cloud, and it attracted the most gas and eventually became the largest planet.

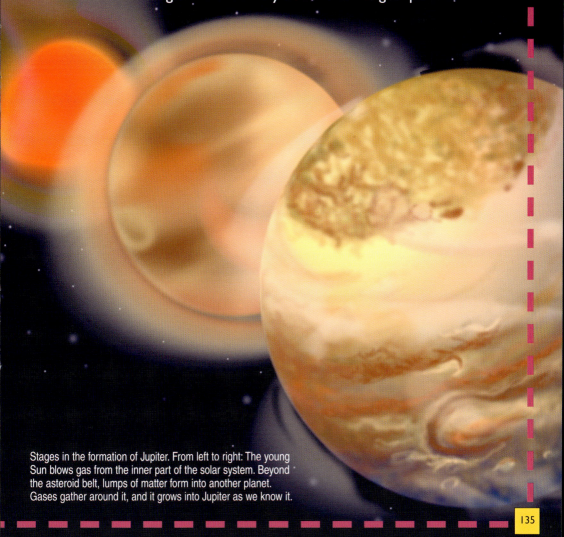

Stages in the formation of Jupiter. From left to right: The young Sun blows gas from the inner part of the solar system. Beyond the asteroid belt, lumps of matter form into another planet. Gases gather around it, and it grows into Jupiter as we know it.

Jupiter Basics

Jupiter is by far the largest body in the solar system, after the Sun. It is so large that it could swallow more than 1,000 bodies the size of Earth.

Jupiter is the fifth planet in the solar system, in order of distance from the Sun. On average, it lies about 484 million miles (780 million km) away. The closest it travels to Earth is about 390 million miles (630 million km) away.

Even at such a great distance, Jupiter can often be seen shining brightly in the night sky. This is mainly because it is so big. With a diameter, or distance across, of 88,800 miles (143,000 km), it is 11 times as big across

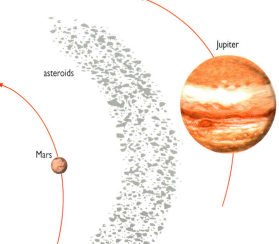

asteroids

Jupiter

Saturn

Mars

Jupiter orbits in the solar system between Mars and Saturn. It lies more than three times farther away from the Sun than Mars.

Above Right: Photographs taken through telescopes on Earth show the multicoloured bands of Jupiter. Various spots also show up, which are the centres of severe storms.

Above: Jupiter is truly a giant of a planet, dwarfing our own planet Earth. Several Earths could fit into Jupiter's huge permanent storm region we know as the Great Red Spot.

STARPOINT

Though Jupiter is the biggest planet, it rotates in space 15 times faster than the smallest planet, Pluto.

as Earth. Its volume is over 1,000 times that of Earth, which means that Jupiter takes up 1,000 times as much space as Earth takes up. Jupiter's mass, or the amount of matter it contains, is about 318 times Earth's mass.

SPINNING AROUND

As Jupiter travels around the Sun, it also moves in another way. Like all the planets, Jupiter rotates, or spins around on its axis. An axis is an imaginary line through an object from its north pole to its south pole.

Jupiter rotates faster than any other planet. In fact, this huge planet spins all the way around once in less than 10 hours. This is less than half the time Earth takes to spin around once, which is 24 hours, or 1 day. Because Jupiter spins so fast, it bulges out at the middle around its equator—the imaginary line around the centre of a planet, midway between its north and south poles. This fast spinning also causes Jupiter to be slightly flattened at its poles.

JUPITER'S MAKEUP

Jupiter's makeup is very different from that of a rocky planet like Earth. Earth is made up of three main layers of rock, with a metal core at the centre. Jupiter is made up of layers too, but mostly of gas and liquid gas.

When we look at Jupiter through a telescope, we can see various coloured markings, including light bands and dark bands. These markings are actually the clouds of Jupiter's atmosphere, which is its top layer. The temperature here is very cold—about –200°F (–130°C).

Below the atmosphere is Jupiter's surface. The planet's surface is not solid, but instead is a vast ocean of liquid hydrogen. In this part of Jupiter, the atmospheric pressure—the force of the gases pressing down—is enormous. Deeper down inside Jupiter, the pressure is so great that it forces the hydrogen to turn into a metallic liquid. Temperatures here may rise as high as 43,000°F (24,000°C). At the centre of the planet is probably a core of rock and ice, with about 10 to 20 times the mass of Earth.

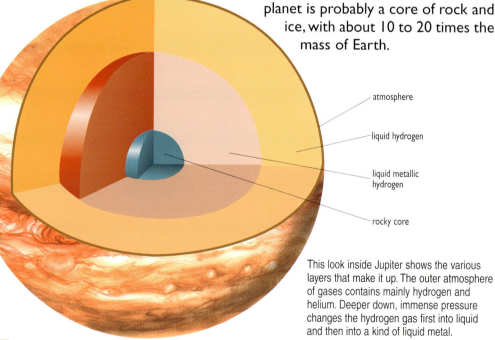

atmosphere

liquid hydrogen

liquid metallic hydrogen

rocky core

This look inside Jupiter shows the various layers that make it up. The outer atmosphere of gases contains mainly hydrogen and helium. Deeper down, immense pressure changes the hydrogen gas first into liquid and then into a kind of liquid metal.

Stormy Atmosphere

Clouds and furious winds rush through Jupiter's atmosphere. Lightning flashes among the clouds, as storms spring up all over the planet.

Hydrogen and helium are the two main gases in Jupiter's stormy atmosphere. Traces of other gases are found there as well. They include ammonia, water vapour (water in the form of a gas), and methane. Methane is the main gas in the natural gas used to heat homes on Earth. Gases containing sulphur are also found in Jupiter's atmosphere.

The coloured bands in Jupiter's atmosphere are actually layers of clouds. The clouds form into bands because the atmosphere moves so quickly. Astronomers call the pale bands zones and the darker bands belts. The main colours in the atmosphere are white and red. The white regions are high clouds of tiny ammonia crystals. The reddish ones are lower clouds made up of sulphur compounds.

Jupiter's main visible features are the dark and light bands, which astronomers call belts and zones. They have each been given names so that astronomers can refer to them when reporting the changes taking place on the planet.

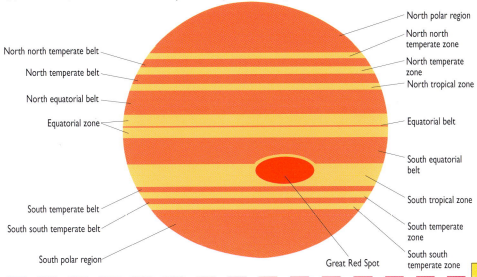

North polar region
North north temperate zone
North temperate zone
North tropical zone
Equatorial belt
South equatorial belt
South tropical zone
South temperate zone
South south temperate zone

North north temperate belt
North temperate belt
North equatorial belt
Equatorial zone
South temperate belt
South south temperate belt
South polar region

Great Red Spot

139

Above: The appearance of the belts and zones on Jupiter changes hour by hour as furious winds rage in the atmosphere.

In the belts and zones, the winds blow strongly, at speeds of up to 300 miles (500 km) an hour. They do not all blow in the same direction. Some blow toward the east, in the same direction the planet rotates. Other winds blow toward the west.

STORMY WEATHER

Where the winds on Jupiter blowing in opposite directions meet, the atmosphere swirls around furiously. Great wavy patterns, or eddies, form, as do white and coloured spots. In these spots, the atmosphere whirls around like in a gigantic whirlwind. Lightning flashes between the clouds as it does in thunderstorms on Earth.

LIFE AMONG THE CLOUDS?

The deeper one goes into Jupiter, the higher the temperature. Beneath the cold upper atmosphere are warm layers of gas. In the warm layers, some scientists think that chemical reactions might take place between the gases of the atmosphere, such as ammonia, methane, and water. Scientists believe that long ago these gases might have combined to form simple living things. This is probably what happened on Earth billions of years ago.

Some of the stormiest regions on Jupiter occur around the Great Red Spot. Here the clouds form fascinating wavy patterns as the winds swirl and eddy.

On Jupiter, say scientists, it is possible that special life forms spend their whole lives floating in the warm layers of the atmosphere.

THE GREAT RED SPOT

The Great Red Spot is Jupiter's biggest feature. Astronomers had noticed this reddish oval region in Jupiter's atmosphere and wondered what it was for more than 300 years. Space probes have shown that the Great Red Spot is an enormous storm, much like a hurricane on Earth.

The Great Red Spot measures about 25,000 miles (40,000 km) across. Three bodies the size of Earth could fit in it side by side. The spot changes its size and colour slightly over time. But its position on Jupiter never changes.

The Great Red Spot spins around once about every six days.

This close-up of the Great Red Spot, was taken by a Voyager space probe. The gases in this great storm region circulate in a counterclockwise direction. The colour of the Spot seems to come from the presence of the chemical element phosphorus.

Jupiter's Magnetism

STARPOINT

Jupiter's magnetism can be detected as far away as the orbit of Saturn, which is 400 million miles (600 million km) away from Jupiter.

Jupiter is a magnetic planet, like Earth. But its magnetism is much more powerful and reaches out millions of miles into space.

All around Earth is an invisible force called magnetism. Earth's magnetism is produced in its core, which is made up mainly of hot liquid iron. When Earth rotates, the movement creates currents of electricity in the core.

Jupiter has magnetism too, and astronomers believe Jupiter's magnetism is produced in much the same way Earth's is—by electric currents in liquid metal. But on Jupiter, the liquid metal is not found in an iron core. The source of Jupiter's magnetism is the metallic hydrogen found in the layer surrounding Jupiter's core.

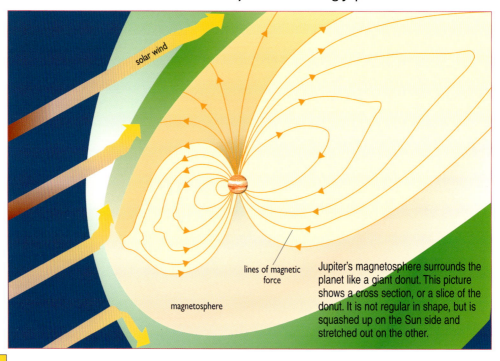

solar wind

lines of magnetic force

magnetosphere

Jupiter's magnetosphere surrounds the planet like a giant donut. This picture shows a cross section, or a slice of the donut. It is not regular in shape, but is squashed up on the Sun side and stretched out on the other.

THE MAGNETOSPHERE

Jupiter's magnetism is 20,000 times as strong as Earth's. It acts all around the planet and reaches out millions of miles into space. The magnetic region is called the magnetosphere. It is constantly moving in space because it rotates with Jupiter.

The magnetosphere does not stretch evenly in all directions around Jupiter. This is because of the solar wind—the stream of particles given off by the Sun.

Jupiter's "northern lights" show up in this picture of the planet taken by the Hubble Space Telescope. They are displays of aurora around the planet's north polar regions.

On the side of Jupiter facing the Sun, the magnetosphere is squashed somewhat by the solar wind.

Some of the particles in the solar wind flow around the magnetosphere. Some get trapped inside it. Others travel in toward Jupiter and enter its atmosphere. They make the gases in the atmosphere glow. Something similar happens on Earth when particles from the solar wind enter Earth's atmosphere and make it glow. We call these shimmering lights the aurora, or the Northern and Southern Lights.

RADIO JUPITER

Unlike other planets, Jupiter gives off strong radio radiation. Astronomers pick up these waves with radio telescopes on Earth. Most astronomers believe the radio radiation is caused by particles from the solar wind whizzing around in Jupiter's magnetosphere. Other radio waves are produced by powerful flashes of lightning during the great storms that often take place in Jupiter's atmosphere.

Powerful Gravity

Jupiter has an extremely powerful gravitational pull. It keeps a large family of moons in place, and it affects other nearby bodies in space.

If you could live on Jupiter, you would find if difficult to walk. This is because Jupiter's gravity is more than two and a half times more powerful than Earth's gravity.

Left: Europa is one of the 16 moons that Jupiter holds on to with its powerful gravity. In the solar system, only the Sun has a stronger gravitational pull than Jupiter.

Below: Jupiter is at the centre of its own miniature "solar system," formed by its 16 circling moons. They fall into four families, according to their orbits. From one side to the other, Jupiter's "solar system" spans a distance of nearly 30 million miles (50 million km).

Galilean moons

Jupiter

outer family

middle family

inner family

That means that if you weigh 100 pounds on Earth, you would weigh 264 pounds on Jupiter.

Every body in the universe has gravity due to its mass. The more a body's mass, the more powerful its gravitational pull is. Jupiter has a powerful pull because it is so big.

Jupiter's gravity reaches out a long way into space. It is the force that keeps Jupiter's 16 moons circling around the planet. Even 15 million miles (25 million km) away, Jupiter's gravity is strong enough to hold on to a tiny rocky moon called Sinope, which is only about 22 miles (35 km) across.

JUPITER AND THE ASTEROIDS
Jupiter's gravity also affects some of the asteroids. Asteroids are rocky bodies that orbit the Sun like small planets. Most asteroids are found between the orbits of Mars and Jupiter. But Jupiter has captured two groups of asteroids called the Trojans. They travel around the Sun in the same orbit as Jupiter. One group travels in front of Jupiter, and the other travels behind it.

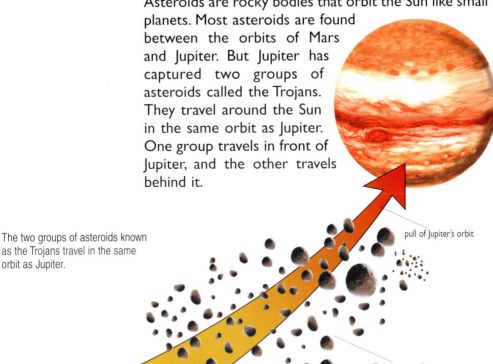

The two groups of asteroids known as the Trojans travel in the same orbit as Jupiter.

pull of Jupiter's orbit

Trojan asteroids

145

Above: Many comets are affected by Jupiter's powerful gravity. When this happens, they may become regular visitors to our skies, like Halley's comet, shown here.

JUPITER AND COMETS

Jupiter's gravity may also affect comets when they travel near the planet. Comets are tiny balls of ice and dust that orbit the Sun. They travel in toward the Sun from the outer parts of the solar system. Afterward, they usually travel back to where they came from. Most comets take thousands of years to make the journey in toward the Sun then back out again.

If a comet passes near Jupiter, the giant planet's gravity pulls it and changes its path. This may force a comet into a much shorter orbit so that it travels back to the Sun in a matter of years rather than thousands of years. Several comets travel between the Sun and Jupiter, and astronomers say that they belong to Jupiter's comet family.

COMET COLLISIONS

Jupiter's powerful gravity sometimes causes a passing comet to break apart. This happened to a comet called Shoemaker-Levy 9 in 1992. In March 1993, astronomers discovered that pieces of the comet were travelling toward Jupiter. In July 1994, the pieces smashed into Jupiter. Some of the collisions created huge fireballs. The collisions left

Top: Many comets are affected by Jupiter's powerful gravity. When this happens, they may become regular visitors to our skies, like Halley's comet, shown here.

Left: In the summer of 1994, a string of fragments of the comet Shoemaker-Levy 9 made a beeline for Jupiter. When they crashed into the planet, they left prominent "scars" in the atmosphere (inset picture).

markings in Jupiter's atmosphere that lasted for days. The largest was almost as big as Earth.

JUPITER'S RINGS

Until 1979, astronomers knew of only two planets—Saturn and Uranus—that had rings around them. But in 1979 the space probe Voyager 1 spotted a system of very thin, faint rings when it flew past Jupiter. This was a complete surprise. In 1984, astronomers discovered that Neptune has rings, too.

The main part of Jupiter's ring is located about 76,500 miles (123,000 km) from the planet, and it circles the planet's equator. The ring is only about 4,000 miles (6,000 km) across and about 19 miles (30 km) thick. Smaller rings circle inside and outside this main ring.

The rings are not solid, like metal washers. They are made up of tiny particles of rock. These particles travel in orbit around Jupiter at high speed. They appear to us as rings because they move so fast that the light they reflect is blurred.

STARPOINT

Particles in Jupiter's rings whiz around the planet in only about six hours.

Part of the ring system around Jupiter, photographed by the Voyager 1 space probe, which discovered it (right). It shows up more clearly in a picture of the planet taken with ultraviolet ligh

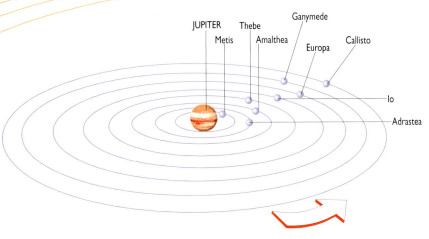

JUPITER
Metis
Thebe
Amalthea
Ganymede
Europa
Callisto
Io
Adrastea

Jupiter's Moons

Jupiter's moons are some of the most fascinating bodies in our solar sytem. Most are tiny, but four are planet-sized.

The Italian astronomer Galileo was first to train a telescope on the heavens and first to spot Jupiter's four large moons.

From Earth, we can observe 13 of Jupiter's 16 known moons. The other three were discovered by the Voyager space probes in 1979 and 1980. Jupiter's four largest moons circle quite close to the planet. They are Io, Europa, Ganymede, and Callisto. These are called the Galilean moons, after the Italian astronomer Galileo. He discovered them with a small homemade telescope in 1610.

The Galilean moons can easily be seen from Earth. Through binoculars, you can see them lined up around Jupiter. But it wasn't until the Voyager probes visited Jupiter in 1979 that we received detailed pictures and

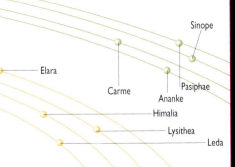

Sinope

Elara

Carme

Pasiphae

Ananke

Himalia

Lysithea

Leda

Right: This montage of Voyager pictures shows Jupiter and the four Galilean moons. From top to bottom, they are Io, Europa, Ganymede, and, in the right-hand corner, Callisto.

information about Jupiter's moons. Astronomers were surprised to learn how much the moons differ from one another.

THE DWARF MOONS

While the Galilean moons measure thousands of miles across, most of Jupiter's other moons are much smaller. The smaller moons range in size from 9 to 106 miles (15 to 170 km) in diameter. Metis, the moon closest to the planet, measures only about 25 miles (40 km) across.

Four of Jupiter's tiny moons circle the planet 10 times farther out than Callisto, the farthest out of the Galilean moons. They circle Jupiter at a distance of more than 7 million miles (11 million km). Another group of four moons circle the planet at about twice this distance. Jupiter's most distant moon, Sinope, takes more than three Earth-years to circle around Jupiter.

Left: The solar system's biggest moon, Ganymede. Bright specks show where meteorites have hit the surface and dug out icy craters. The brighter regions on the moon are thought to be younger than the darker ones.

GANYMEDE

Ganymede is Jupiter's largest moon. It is also the largest moon in the entire solar system. With a diameter of 3,278 miles (5,276 km), it is two and a half times as large as our own Moon, and it is larger than the planet Mercury.

Scientists believe that Ganymede has three layers. In its centre is a small core of iron or iron and sulphur. Surrounding the core is a rocky mantle, and on top is an icy surface layer, or crust. Ganymede's surface has dark rocky areas and white icy areas. Craters created by meteorites cover much of the dark areas. Meteorites are lumps of rock that fall from space onto a planet or moon's surface.

Ganymede's icy areas are covered by dark grooves, or valleys, which lie between long, narrow ridges. Astronomers believe these grooves and ridges were caused by movements in Ganymede's crust long ago.

Above: Callisto is the most heavily cratered of Jupiter's moons. It probably has the oldest surface.

Ice-covered Europa is the smoothest body we know in the solar system.

Above: Few craters are visible on the smooth surface of Europa. It is the smallest of the Galilean moons, with a diameter of 1,942 miles (3,126 km).

Right: Europa has a network of dark lines, which are probably cracks in the icy surface that have filled with dirty material from below.

CALLISTO

Callisto is Jupiter's second-largest moon. Measuring 2,995 miles (4,820 km) across, it is nearly the same size as the planet Mercury. Like Ganymede, it is also made up of rock and ice.

But unlike Ganymede, Callisto is completely covered with craters. It has more craters than any other moon in the solar system. Most of these craters were formed billions of years ago. Callisto has no mountains and valleys like those found on Ganymede. So astronomers believe that there have been no movements in Callisto's crust for billions of years.

EUROPA

Europa looks very different from the other Galilean moons. Completely covered with ice, Europa's surface is light in colour and very smooth. The icy surface probably formed when liquid water seeped up from below and froze.

A network of dark grooves and ridges can be seen all over Europa. The dark grooves are probably cracks in the icy crust. They extend up to thousands of miles across the surface. Scientists believe that an ocean of liquid water may still lie under the icy surface.

Io, the Pizza Moon

Io is the Galilean moon that is closest to Jupiter, and it is a very unusual moon. Io is the only body in the solar system, other than Earth, with active volcanoes. Its brightly coloured markings are caused by volcanoes erupting on its surface.

With a diameter of 2,255 miles (3,630 km), Io is slightly larger than our own Moon. While most other moons are dull in colour, Io is mainly yellow and orange, with pale and dark markings. It has been called the pizza moon because its surface looks somewhat like the colourful surface of a pizza.

The astonishing surface of Io (top) owes its colour to the sulphur that spews out of its many volcanoes. Vapour and dust erupting from the volcanoes shoot high above the surface (below).

Sulphur Landscape

The volcanoes on Io are not like the volcanoes on Earth. On Earth, volcanoes are places where red-hot liquid rock under the ground forces its way up to the surface. The rock then quickly cools and becomes hard.

On Io, the substance that forces its way to the surface in volcanoes is sulphur. Sulphur is yellowish-orange in colour, which explains the colour of Io's surface. Sulphur is found around volcanoes on Earth in small amounts. On Io, sulphur from volcanoes has covered the whole surface.

Io's volcanoes also shoot large amounts of sulphur vapour into the sky. This freezes to form particles of a white substance that falls like snow on the surrounding landscape.

HOW THE VOLCANOES FORM

The volcanoes on Io are almost certainly caused by Jupiter's enormous gravity. It pulls different parts of Io at different times as the moon spins around in space. This sets up movements inside Io, which make it heat up. The heat melts the sulphur and forces it out through cracks in the hard outer crust.

In this Voyager picture, Io is seen over Jupiter's famous Great Red Spot. The other moon in the picture is Europa.

Linda's Luck

The first volcano on Io was discovered by Linda Morabito of the Jet Propulsion Laboratory in California. She found it in February 1979 when studying photographs of Jupiter and its moons taken by the Voyager 1 space probe.

Missions to Jupiter

In the 1970s space probes from NASA gave us our first close look at Jupiter and its moons. They sent back amazing pictures and vast quantities of new information.

In March 1972 a powerful Atlas-Centaur rocket blasted off the launch pad at Cape Canaveral in Florida. It was carrying a space probe called Pioneer 10. Faster and faster the rocket sped into the air and then into space. When its engines stopped firing, the space probe separated and set off on a journey that would take it to Jupiter. The probe left Earth travelling at a speed of more than 32,000 miles (51,000 km) an hour. This was

Pioneer 10 was a true pioneer of deep space exploration, becoming the first space probe to navigate the asteroid belt and experience the intense radiation in Jupiter's magnetosphere. Pioneer 11 followed a different path around Jupiter so that it could carry on to Saturn.

the fastest any vehicle had ever travelled—more than 15 times faster than a rifle bullet.

The biggest part of Pioneer 10 was the dish antenna, which measured about 9 feet (2.7 m) across. Behind it was the main body of the spacecraft, which included a box containing electronic equipment and various information-gathering instruments.

DODGING THE ASTEROIDS

To reach Jupiter, Pioneer 10 had to pass through the asteroid belt. In this region, thousands of asteroids circle the Sun like miniature planets. No one knew for sure whether the spacecraft could get through the belt without being hit. But it did, and it flew past Jupiter in December 1973.

As it passed by, Pioneer took photographs of Jupiter. They showed that the Great Red Spot is a huge weather system. Pioneer's instruments showed that Jupiter has a powerful magnetic field and gives off radiation. Scientists also learned that the planet gives off more heat than it receives from the Sun.

REPEAT PERFORMANCE

After travelling past Jupiter, Pioneer 10 sped off into interplanetary space. But it continued to send back infor-mation. Pioneer 11 was launched in April 1973 and it reached Jupiter in December 1974. It zoomed in to photograph the planet's south pole. Then, after swinging around Jupiter, it began to follow a long, looping path that would take it to Saturn.

Pioneer 11 showed in detail the vigorous activity taking place in Jupiter's atmosphere for the first time. This view shows the planet's north polar region.

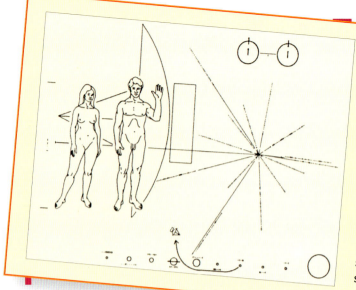

Pictures for Aliens

After leaving the solar system, the Pioneer probes will venture into interstellar space—the space between the stars. One day, perhaps, intelligent aliens somewhere in space may come across them and wonder where they came from. To give them an answer, each Pioneer spacecraft carries a metal plaque shown here with a message in pictures. It tells where the spacecraft came from and who sent it.

DEEP SPACE VOYAGERS

The Pioneer probes proved that spacecraft could safely travel through the asteroid belt and send pictures and information over vast distances. So NASA decided to send two more probes to Jupiter and beyond. Named Voyager 1 and Voyager 2, they were launched in the summer of 1977.

The Voyagers had a dish antenna about 12 feet (3.7 m) across. Most of its instruments were carried on a movable platform so that they could easily be pointed in different directions. Its cameras were much better than those on the Pioneers.

ASTOUNDING IMAGES

Voyager 1 made its closest approach to Jupiter in March 1979. But long before that it was sending back pictures that were astounding mission scientists. They showed the swirling atmosphere of the giant planet in great detail. Scientists saw clouds racing along at high speeds and storms breaking out all over Jupiter.

Voyager also sent back pictures of Jupiter's moons.

Voyager 2 blasts off the launch pad at Cape Canaveral on August 20, 1977, about two weeks before its sister craft, Voyager 1. But Voyager 1 reached Jupiter first.

Every moon they spied looked different, which came as a big surprise. Oddest of all was the brilliantly coloured Io, nicknamed the pizza moon, which had volcanoes. Then came another surprise, when Voyager 1 spotted Jupiter's ring.

Soon Voyager 1 was moving on to its next target, Saturn. Voyager 2 took its place, flying closest to the planet in July 1979. It sent back more information and more amazing pictures of storms, moons, volcanoes, and rings. In just a few months, the two Voyager spacecraft had revolutionized our knowledge of the biggest planet in our solar system.

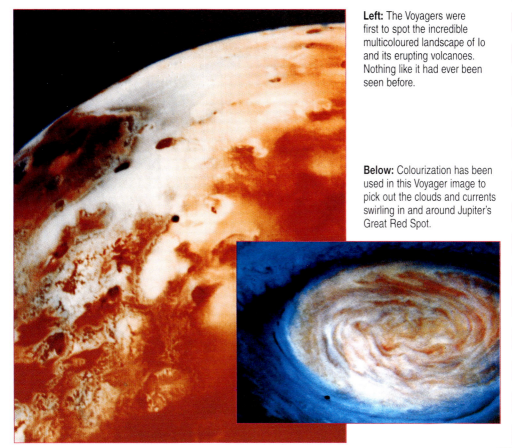

Left: The Voyagers were first to spot the incredible multicoloured landscape of Io and its erupting volcanoes. Nothing like it had ever been seen before.

Below: Colourization has been used in this Voyager image to pick out the clouds and currents swirling in and around Jupiter's Great Red Spot.

Saturn

Saturn is the second largest planet in the solar system—the family of bodies that circle in space around the Sun. Saturn's most unique feature is the bright set of rings that travel around it. Three other planets have rings, but only Saturn's rings can be seen from Earth. These rings make Saturn one of the most beautiful bodies in the solar system.

Saturn, along with Jupiter, Uranus, and Neptune, is one of the gas giants. Unlike Earth, Mercury, Mars, and Venus, which are made up mainly of rock, Saturn is made up mainly of gas. Its makeup causes it to be extremely light for its size. In fact, it is the lightest planet in the solar system. If you could put it in water, it would float. Every other planet would sink.

Like all the gas giants, Saturn travels through space with a large family of moons. Of all the planets, Saturn has the largest family of known moons—at least 18. One of them, Titan, is larger than the planet Mercury. Titan is also the only moon in the solar system that has a thick atmosphere, or a layer of gases around it.

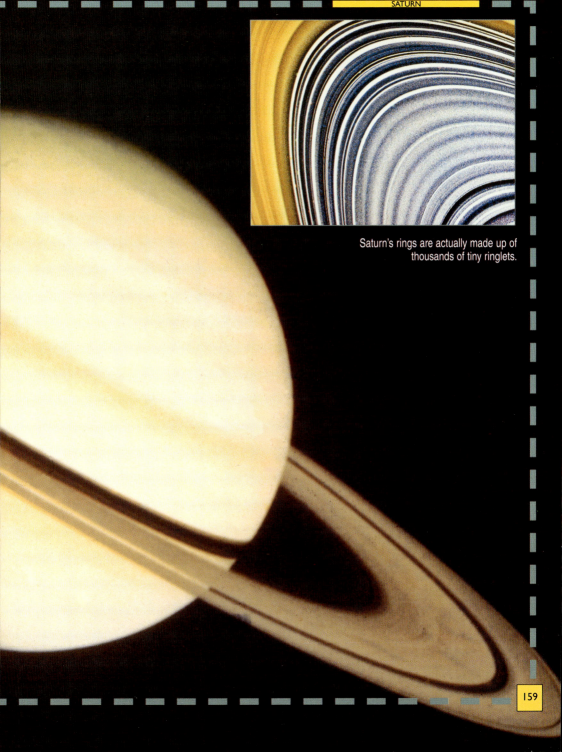

Saturn's rings are actually made up of
thousands of tiny ringlets.

Saturn Basics

Saturn is a rapidly spinning planet, made special by the rings that travel around it.

Saturn is the sixth planet in the solar system in order going out from the Sun. On average, it lies about 877 million miles (1.4 billion km)

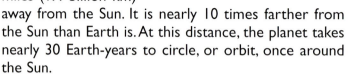

Saturn is so large that it could swallow nearly 750 bodies the size of Earth.

away from the Sun. It is nearly 10 times farther from the Sun than Earth is. At this distance, the planet takes nearly 30 Earth-years to circle, or orbit, once around the Sun.

Because it is so remote, Saturn is not an easy planet to spot in the night sky. During most of the year, it does not shine as brightly as Venus, Jupiter, or Mars, which are all closer to the Sun. But at its brightest, Saturn can outshine most of the stars in the sky. Like all the planets, Saturn does not shine by its own light. It shines because it reflects light from the Sun.

THROUGH A TELESCOPE
Through a small telescope, Saturn looks different from

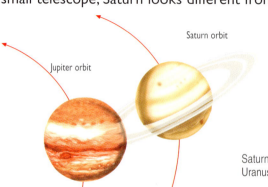

Jupiter orbit

Saturn orbit

Uranus orbit

Saturn orbits between Jupiter and Uranus. All three planets are gas giants

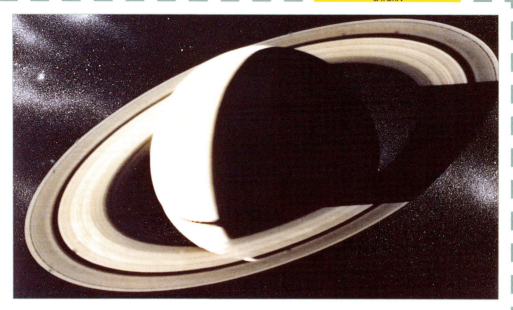

the other planets. The planet appears to be oval shaped rather than spherical, or round. A more powerful telescope will show why—the planet is surrounded by rings. The edges of the rings on either side make the image look oval.

In this close-up view of Saturn, we can see how the planet casts a shadow on its rings.

Saturn appears to us to be yellow. This yellow colouring is not the planet's surface but its thick atmosphere. The main marking we can see on Saturn's yellow body is the black shadow cast by the rings. Occasionally we can spot small black dots, which are the shadows of Saturn's larger moons. Faint parallel bands can also be seen on Saturn. Astronomers call the darker bands belts and the lighter ones zones. Other faint dark and light streaks and spots appear on Saturn from time to time.

How Saturn Formed

Saturn was formed along with the other planets about 4.6 billion years ago. It formed from gases and icy chunks of rock that came together in the outer solar system. Over time, the rocky materials formed into a larger and larger ball, which astronomers believe grew to become a body at least the size of Earth. Then the rocky body attracted gases around it until it became the gas giant we call Saturn.

As Saturn travels in its orbit, different parts of the planet lean closer to the Sun and receive more of its warmth.

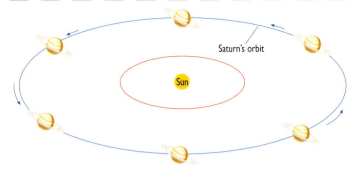

Saturn's orbit

Sun

IN A SPIN

Saturn takes a little over 10½ hours to rotate, or spin around, once. This is less than half the time that Earth takes to complete one rotation, which is 24 hours, or one day. In fact, except for Jupiter, Saturn rotates more quickly than any other planet in the solar system.

Because Saturn rotates so quickly, it bulges out at the equator and is flattened at the poles. A planet's equator is the imaginary line around it, midway between its north and south poles. This flattening also occurs on Earth, but on Saturn, the flattening and bulging effects are much greater. That's because the planet is made up mainly of gas and liquid, which can change shape easily. This makes Saturn's diameter, or distance across, nearly 7,500 miles (12,000 km) greater at the equator than at the poles.

SATURN'S TILT

Like all planets, Saturn rotates on its axis—an imaginary line running through a planet from its north pole to its south pole. In some planets, the axis is almost vertical, or upright, in relation to the planet's orbit. But in Saturn, the axis is tilted at an angle in space. As the planet travels in its orbit, this tilt causes different parts of the planet to lean closer to the Sun. The same thing occurs on Earth and brings about the regular changes in the weather we call the

Saturn's Warmth

Given Saturn's great distance from the Sun, temperatures on the planet are much higher than one would expect. Scientists believe this may be partially caused by droplets of liquid helium that form beneath the weight of Saturn's atmosphere. Because liquid helium is much heavier than liquid hydrogen, the droplets sink through Saturn's hydrogen ocean. As they fall, they give off heat. Over millions of years, this effect has caused temperatures on the planet to rise. Temperatures below Saturn's atmosphere may be at least 10,000°F (5,500°C), but its upper atmosphere is only about −300°F (−185°C).

seasons. Because Saturn's orbit takes nearly 30 Earth-years, its seasons last for 7½ years.

Saturn's tilted axis also affects the way the planet appears from Earth. As the planet circles the Sun, we can see different views of the rings. When the planet is tilted toward us, we get our best views of the rings. At other times, the rings seem to almost disappear.

INSIDE SATURN

Like its neighbour Jupiter, Saturn is made up mostly of gas. Unlike Earth, it does not have a solid surface that you could stand on. Beneath a heavy atmosphere of mostly hydrogen gas, great pressure turns the hydrogen into liquid. Astronomers believe that a vast liquid hydrogen ocean, tens of thousands of miles deep, covers the whole planet. Below this hydrogen ocean, even higher pressures turn the hydrogen into liquid metal. This liquid metal probably surrounds a small core of rock and ice at least the size of Earth.

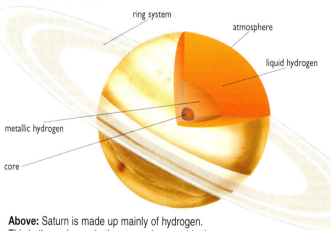

ring system
atmosphere
liquid hydrogen
metallic hydrogen
core

Above: Saturn is made up mainly of hydrogen. This is the main gas in the atmosphere and in the two layers that surround Saturn's core.

Right: Saturn's tilted position in space causes us to see different views of its rings from Earth. The diagram shows how our view of them will change over time.

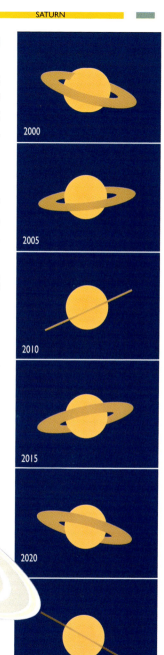

2000

2005

2010

2015

2020

2025

SATURN DATA

Diameter at equator:
74,600 miles (120,000 km)

Average distance from Sun:
887,000,000 miles
(1,430,000,000 km)

Rotates in:
10 hours, 39 minutes

Orbits Sun in: 29.5 years

Moons: 18 known

SATURN'S MAGNETISM

Like Earth and Jupiter, Saturn has a liquid metal layer that causes it to be magnetic. As Saturn rotates, the movement creates currents of electricity in its metal layer. These currents produce Saturn's magnetism. This is similar to the way Earth produces its magnetism. On Earth, magnetism is the force that makes a compass needle point north.

Saturn's magnetism extends many millions of miles out into space around the planet, forming a magnetic bubble known as the magnetosphere. A stream of particles from the Sun called the solar wind flows around the magnetosphere. Some of the particles in the solar wind get trapped in certain parts of Saturn's magnetosphere. The Earth's magnetosphere has similar areas, called the Van Allen belts, which produce powerful radiation. Other solar wind particles enter Saturn's atmosphere and make it glow. These light displays, or aurora, are similar to the Northern and Southern Lights we sometimes see on Earth.

Saturn's magnetism stretches far out into space. It acts as a barrier to the solar wind, and causes it to change direction.

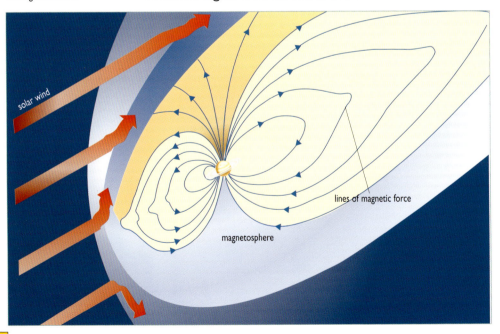

solar wind

lines of magnetic force

magnetosphere

Saturn's Atmosphere

Strong winds blow in Saturn's atmosphere, causing clouds to swirl around the planet and creating fierce storms.

Hydrogen makes up about 94 percent of Saturn's atmosphere, and helium makes up most of the rest. There are also traces of other gases, including methane, ammonia, and water vapour. These gases form the cloud features that appear as the faint bands called belts and zones that we see on Saturn. Partly hidden by a thick haze in the upper atmosphere, these clouds create Saturn's pale yellow colour.

Temperatures in Saturn's upper atmosphere are very low, and the clouds there are made up of frozen gas crystals. Beneath these high clouds are clouds formed from water droplets. This lower layer of clouds is similar to clouds in Earth's atmosphere.

This colourised picture of Saturn shows the bands of clouds in the planet's atmosphere. Without colourisation, the parallel bands are difficult to see.

WILD WEATHER

In Saturn's atmosphere, winds blow violently, furious storms rage, and lightning flashes. On Earth, heat from the Sun produces our weather. But on Saturn, heat from inside the planet creates most of the planet's weather.

In this prominent band in Saturn's northern hemisphere, winds travel at speeds approaching 300 miles (500 km) an hour.

POWERFUL JETS

Saturn's winds blow in powerful jet streams, or fast-moving air currents in the atmosphere. The most powerful one is the broad equatorial jet stream that extends for thousands of miles on either side of Saturn's equator. Winds within this jet stream whip around the planet at speeds of up to 1,100 miles (1,800 km) an hour. This is more than three times as strong as the winds in Earth's fiercest tornadoes.

The equatorial jet stream blows toward the east, in the same direction as Saturn's rotation. Farther north and south of the equator are several narrower jet streams that blow in the opposite direction.

The oval regions in this colourized picture of Saturn's atmosphere are violent storms.

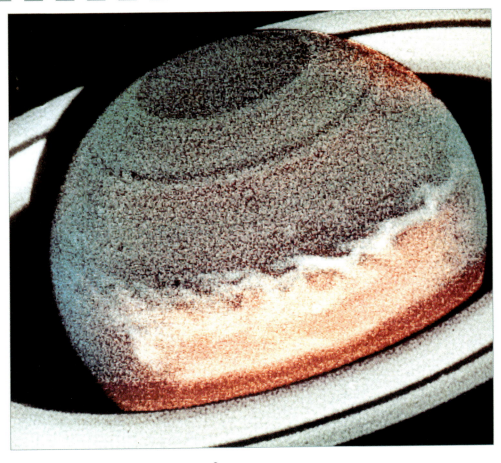

This Hubble Space Telescope picture shows Saturn streaked by violent weather and high winds that reach more than 1,100 miles (1,800 km) per hour.

SPOTTING STORMS

Violent storms occur at the edges of Saturn's jet streams, where the atmosphere gets churned up. Storms also occur within the jet streams themselves, caused by warm air circulating upward from lower levels in the atmosphere. From Earth, large storm areas in Saturn's atmosphere often appear as oval patches or spots. These spots can look white, brown, or red. They are similar to the spots that appear on Jupiter but are not as striking or as big. Nothing on Saturn is like the huge storm on Jupiter known as the Great Red Spot, which has been seen on the planet for centuries.

The Remarkable Rings

Four planets in the solar system have rings, but only Saturn's large rings are bright enough to be viewed through telescopes on Earth.

When the Italian astronomer Galileo discovered Saturn in 1610, he noticed that it looked different from the other planets he had observed with his telescope. Other planets appeared to be round, but Saturn seemed to have ears, or strange attachments on each side. Galileo thought they might be moons. His telescope was not powerful enough to show what the "ears" really were—the outer edges of a huge system of rings.

Above: This superb picture of Saturn's rings shows the many ringlets that make up Saturn's main ring system.

Below: Another view of Saturn's rings clearly shows the thin F ring at the top.

Saturn's rings circle around the planet's equator. From edge to edge, the ring system measures more than 250,000 miles (400,000 km) wide. On Earth, we can see Saturn's three main rings. Astronomers have labeled them A, B, and C from the outside going in. From edge to edge, the A, B, and C rings appear to be about 40,000 miles (64,000 km) wide.

SHINING BRIGHT

Saturn's rings reflect sunlight brilliantly. Without its rings, the planet would appear much fainter in the night sky. We know this because at two points in Saturn's 30-year journey around the Sun, the rings almost disappear from our view. When this happens, Saturn fades noticeably.

This picture of saturn's rings is unusual because the normally bright B ring is not visible.

The three rings vary widely in their individual brightness. The middle B ring is the brightest, followed by the outer A ring. The inner C ring is very faint and transparent. The body of the planet can be seen through it.

MORE RINGS

Along with Saturn's three major rings, astronomers have discovered several other rings. A faint ring, which scientists have labeled the D ring, circles close to Saturn inside the C ring. This ring is so close to Saturn that its inner edge probably touches the planet's atmosphere. More rings, which scientists have labeled F, G, and E, circle Saturn outside the A ring. The F and G rings are narrow, while the E ring is wide but very faint. This E ring is so far from Saturn that it travels between the orbits of Saturn's inner moons, Mimas and Enceladus.

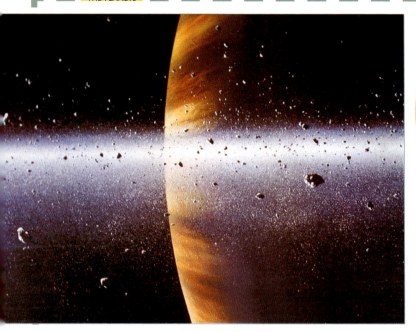

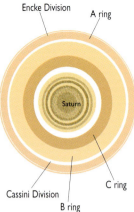

Encke Division

A ring

Saturn

Cassini Division

B ring

C ring

Above: This diagram shows Saturn's main ring system, as seen from a point above the planet's north pole.

Left: Saturn's rings are actually made up of many small particles that circle the planet at high speed.

RACING RINGLETS

Astronomers have known for centuries that Saturn's rings could not be made up of one solid sheet of matter. Close-up pictures taken by space probes have shown that each ring is actually made up of hundreds or thousands of narrower ringlets. These ringlets are made of moving objects that vary in size from fine dust-like particles to chunks as big as trucks. Scientists think that the particles are mostly icy lumps and some ice-coated rocks. They look like rings because the light they reflect is blurred as they whiz around the planet.

RING GAPS

The rings we see from Earth do not form one continuous set of rings. In 1672, the Italian-French astronomer Giovanni Domenico Cassini spotted a gap between the A and B rings. It became known as the Cassini Division. More than 150 years later, the German astronomer Johann Encke noticed a narrower gap inside the A ring, known as the Encke Division. From Earth, the Cassini and Encke Divisions appear to be empty gaps. However, they actually contain some ringlets. Many other smaller gaps exist where there are fewer ringlets than usual.

Scientists are not sure why these gaps exist. They may be caused by the gravity, or pull, of nearby moons, such as Mimas. This moon's gravity may pull particles away from certain parts of the rings.

This close view of Saturn's A ring shows variations in its ringlets. The dark gap in the rings is the Encke Division.

WHERE THE RINGS CAME FROM

For many years, astronomers believed that Saturn's rings had formed from the remains of bodies that strayed close to the giant planet. If a smaller body, such as a moon, travelled close to Saturn, the massive planet's gravity could have eventually pulled the body to pieces.

However, the more we learn about the rings, the more mysterious they become. Astronomers are no longer certain exactly where Saturn's large ring system came from. Some scientists believe that the rings we see cannot be the remains of smaller bodies broken up and scattered long ago. Instead, Saturn's ring system may be continually forming and reforming over time.

The Spokes
Saturn's rings have an interesting feature that cannot be seen from Earth. Astronomers have discovered that the B ring has spokes. These markings are dark lines fanning out across the B ring, similar to the spokes of a wheel. Up to thousands of miles long, they often appear and disappear within a few hours. Astronomers think the spokes might be caused by fine dust particles that lift up above the rings.

STARPOINT
Saturn's rings are wide but very thin. In certain parts, they are probably less than 100 feet (30 m) thick from top to bottom.

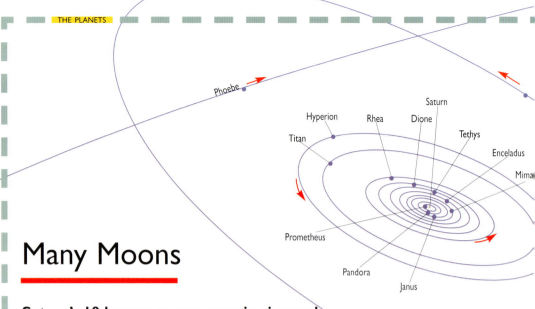

Phoebe

Hyperion · Rhea · Dione · Saturn

Titan · Tethys · Enceladus · Mima

Prometheus

Pandora

Janus

Many Moons

Saturn's 18 known moons vary in size and shape, from rocky lumps just a few tens of miles across to round bodies the size of planets.

Before space travel was possible, astronomers believed that Saturn had only 11 moons—the ones they could see through telescopes. But space probes allowed astronomers to discover that the planet has at least 18 moons. Other possible moons have been reported, so Saturn may have more moons.

Most of Saturn's moons lie relatively close to the planet. Fourteen moons orbit within a distance of about 330,000 miles (530,000 km) from the planet. They form a kind of inner moon system. Several other moons orbit beyond a vast gap in the moon system, at great distances from the planet.

Like our own Moon, most of Saturn's moons complete one rotation in the same amount of time they take to orbit Saturn. As a result, these moons always keep the same side facing the planet. Astronomers call this a captured rotation. However, the outermost of Saturn's moons, Phoebe, does not have a captured rotation.

Most of Saturn's moons lie relatively close to the planet, but several moons orbit millions of miles away.

THE SHEPHERD MOONS

Of Saturn's smallest moons, some of the most interesting are those that orbit close to the rings. Two moons, Pan and Atlas, orbit close to the edge of the A ring. Like many of Saturn's small moons, they are rocky lumps with an irregular shape. Atlas is about 19 miles (30 km) across, and Pan is even smaller. Just beyond Atlas is another pair of somewhat larger moons— Prometheus, which has a diameter of about 62 miles (100 km), and Pandora, which is about 55 miles (90 km) across. Prometheus orbits just inside the F ring, and Pandora travels just outside of it.

Astronomers believe that these tiny moons play an important part in keeping particles within the nearby rings. They are called shepherd moons because of the way they seem to "herd" the ring particles. As these moons travel around either side of the ring, their gravity helps pull back any particles that stray outside the rings.

This montage of photographs shows Saturn and some of its largest moons. Clockwise from top right, the moons are Titan, Mimas, Tethys, Dione, Enceladus, and Rhea.

IN THE SAME ORBIT

The next small moons beyond Pandora are called Epimetheus and Janus. Both moons are less than 125 miles (200 km) across. These moons are so close together that they seem to travel in the same orbit. Astronomers call them co-orbitals. In fact, they do travel in slightly different orbits at slightly different speeds. The inner moon, Epimetheus, travels slightly faster than the outer moon, Janus. Every four years, when Epimetheus catches up with Janus, the gravitational forces between the two moons causes them to swap orbits. They swap back again the next time they meet.

A number of Saturn's small moons share orbits with larger moons. Telesto and Calypso share an orbit with a larger moon, Tethys. And Helene shares an orbit with another larger moon, Dione. The three small co-orbitals are only about 19 miles (30 km) in diameter.

Below are some of the tiny moons that orbit Saturn. Some are less than 20 miles (30 km) across.

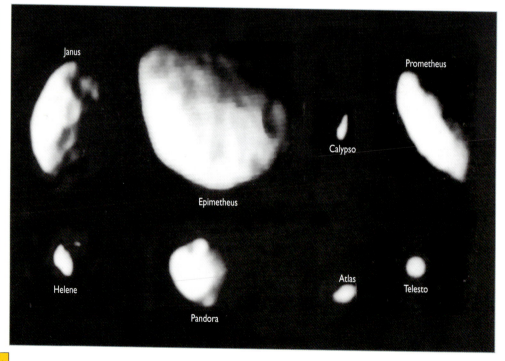

SMALL AND DISTANT

Two small moons orbit at a great distance from Saturn. Hyperion, with a diameter of 180 miles (290 km), is nearly 1.5 million miles (2.4 million km) from Saturn. About the same size as Hyperion, Phoebe is much more remote—nearly 13 million miles (21 million km) from Saturn. This moon takes over 550 days to circle around the planet. That's longer than Earth takes to circle the Sun!

Phoebe is the smallest of Saturn's moons that can be seen through a telescope. It is fairly dark with an irregular shape. Phoebe travels in its orbit in the opposite direction of Saturn's other moons. This suggests that it was not originally part of Saturn's family but was a small body that once orbited the Sun. It was probably captured by the planet's gravity long after the other moons began orbiting Saturn.

ICEBALL MOONS

Saturn's larger moons are spherical, and all but Titan are covered with ice. Five of these round icy moons orbit relatively close to Saturn, beyond the ring system. In order going out from the planet, they are Mimas, Enceladus, Tethys, Dione, and Rhea. The sixth iceball moon, Iapetus, orbits between Hyperion and Phoebe.

All these moons seem to be made up of a mixture of water ice and rock. Beneath their icy surface, some of the larger moons may have a rocky core.

Like Saturn's other large moons, Dione is covered with craters. The largest one in this picture is about 60 miles (100 km) across.

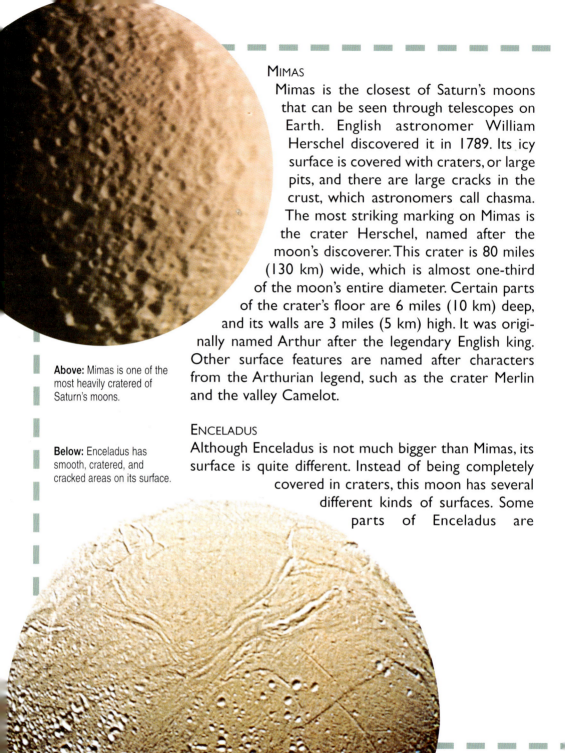

MIMAS

Mimas is the closest of Saturn's moons that can be seen through telescopes on Earth. English astronomer William Herschel discovered it in 1789. Its icy surface is covered with craters, or large pits, and there are large cracks in the crust, which astronomers call chasma. The most striking marking on Mimas is the crater Herschel, named after the moon's discoverer. This crater is 80 miles (130 km) wide, which is almost one-third of the moon's entire diameter. Certain parts of the crater's floor are 6 miles (10 km) deep, and its walls are 3 miles (5 km) high. It was originally named Arthur after the legendary English king. Other surface features are named after characters from the Arthurian legend, such as the crater Merlin and the valley Camelot.

Above: Mimas is one of the most heavily cratered of Saturn's moons.

Below: Enceladus has smooth, cratered, and cracked areas on its surface.

ENCELADUS

Although Enceladus is not much bigger than Mimas, its surface is quite different. Instead of being completely covered in craters, this moon has several different kinds of surfaces. Some parts of Enceladus are cratered, some are covered by long grooves and cracks, and some parts are smooth, with few features at all. These smooth areas are the most mysterious. Astronomers think they were probably created by ice that melted and froze over again, but no one knows exactly what caused the ice to melt.

TETHYS

Astronomers think that this moon must be made up almost completely of water ice, with hardly any rocky material at all. Like Mimas, it is heavily cratered. One crater, Odysseus, is more than 250 miles (400 km) across. It is so big that Mimas could fit into it. Tethys's other remarkable feature is a huge fault, or crack, in its crust, named Ithaca Chasma. In some places the fault is several miles deep and more than 60 miles (100 km) wide. It stretches for more than 1,200 miles (2,000 km). The only other feature we know like it in the solar system is Valles Marineris on Mars.

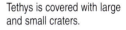

Tethys is covered with large and small craters.

DIONE

Although it is nearly the same size as Tethys, Dione is much heavier, which means it contains more rocky material. Dione is somewhat darker on one side than the other and has bright wispy streaks. These streaks might be caused by fresh icy material that has forced its way to the surface through cracks in the crust. There are also several large craters on Dione, up to 150 miles (240 km) across and with central mountain peaks. Most of Dione's craters are small.

STARPOINT

Enceladus is the most reflective body in the solar system. It reflects nearly all the light that reaches it. In contrast, our own Moon reflects only about 7 percent of the light it receives.

In this photo of Saturn and two of its moons, Tethys and Dione, Tethys is making a shadow on the planet, just below the rings on the far right.

RHEA

Rhea is Saturn's second largest moon. It appears to be made up of equal amounts of ice and rock and probably has a rocky core. Although Rhea is somewhat larger than Dione, it looks very similar. In particular, one side is slightly darker than the other, and it has similar bright wispy markings. Also like Dione, Rhea is heavily cratered, with cracks in some parts of its surface.

IAPETUS

Only slightly smaller than Rhea, Iapetus lies at a great distance from Saturn. At about 2.2 million miles (3.5 million km) away, it orbits among Saturn's most distant moons. Iapetus is one of the more mysterious of Saturn's moons. Even from Earth, astronomers notice that at times the moon is very bright, while at other times it almost disappears. This is because one side of the moon is much brighter than the other. The bright side seems to have an icy, cratered surface, like most of Saturn's moons. But the other side is coated with a dark material that does not reflect much light. Astronomers are not sure what the dark coating is. It may be the result of inner materials that have welled up to the surface.

Unlike Saturn's large moons, small moons like Hyperion have an irregular shape.

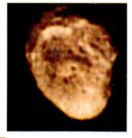

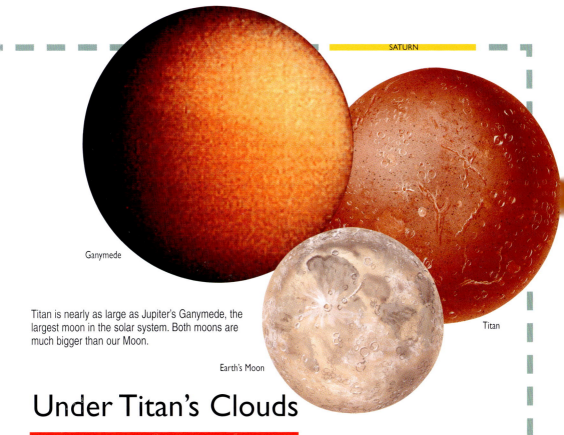

Ganymede

Titan is nearly as large as Jupiter's Ganymede, the largest moon in the solar system. Both moons are much bigger than our Moon.

Earth's Moon

Titan

Under Titan's Clouds

Titan is Saturn's largest moon and the second largest moon in the solar system. Its most unusual feature is its thick atmosphere.

When Dutch astronomer Christiaan Huygens discovered Titan in 1655, other moons around Saturn had not been discovered. Because Titan is so large, it is easy to spot, even with binoculars. With a diameter of 3,200 miles (5,150 km), it is larger than the planet Mercury but smaller than Mars. Among the moons in the solar system, only Jupiter's Ganymede is bigger. Titan orbits Saturn at an average distance of about 750,000 miles (1,200,000 km) and lies relatively close to Hyperion, one of Saturn's outermost moons. Titan takes about 16 days to circle Saturn.

TITAN'S ATMOSPHERE

Titan's most remarkable feature is its thick atmosphere—no other moon in the solar system has one like it. The orange atmosphere, with its deep layers of clouds and a smog-like haze, prevents us from seeing the moon's surface.

Nitrogen, the main gas found on Earth, makes up most of Titan's atmosphere. Other gases, such as methane, hydrogen, and argon, are also present, but only in very small amounts. Because Titan contains a mixture of gases similar to some of those found on Earth, scientists have wondered if Titan could support life. But as far as we know, temperatures on Titan are much too low for any forms of life to exist.

An artist's impression of Titan's surface, with Saturn in the distance

Pioneer Saturn turned its instruments and cameras on Saturn in September 1979. It provided us with much new information about the planet.

Missions to Saturn

Pioneer and Voyager space probes have sent back huge amounts of new information about the mysterious ringed planet.

In April 1973, NASA launched the Pioneer 11 space probe toward Jupiter. It was following in the footsteps of the identical Pioneer 10, launched the previous year. After its encounter with Jupiter in December 1973, Pioneer 10 began heading out of the solar system. But Pioneer 11, after encountering Jupiter in December 1974, headed for an encounter with Saturn. It was renamed Pioneer Saturn.

To reach Saturn from Jupiter, the spacecraft had to travel to the other side of the solar system. The journey took nearly five years. In September 1979, Pioneer Saturn made its closest approach to Saturn when it came within 13,000 miles (21,000 km) of the planet.

Pioneer Saturn sent back the most detailed and close-up pictures of the planet ever seen. It allowed scientists to discover the F ring just outside the A ring, and a new moon, which mission astronomers nicknamed Pioneer Rock.

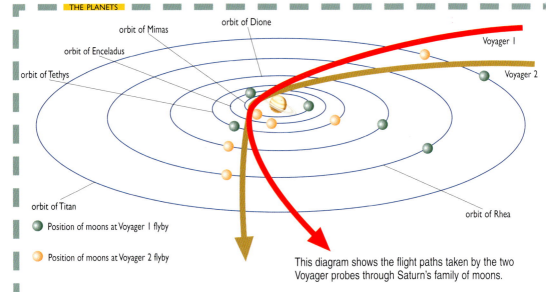

orbit of Mimas
orbit of Dione
orbit of Enceladus
orbit of Tethys
Voyager I
Voyager 2
orbit of Titan
orbit of Rhea

⬤ Position of moons at Voyager I flyby

⬤ Position of moons at Voyager 2 flyby

This diagram shows the flight paths taken by the two Voyager probes through Saturn's family of moons.

THE VOYAGER INVASION

Pioneer Saturn was a pathfinding mission for the next probes to make their way to the ringed planet. They were Voyager 1 and Voyager 2. Like Pioneer Saturn, the Voyagers visited Jupiter first, where they made astounding discoveries. Mission scientists were hoping for similar success when the probes encountered Saturn. They were not disappointed.

REVEALING IMAGES

During October 1980, as Voyager 1 homed in on its target, it began revealing Saturn as never before. Clouds and spots showed up in the atmosphere. New moons were discovered. The rings proved to be made up of thousands of separate ringlets. And the mysterious spokes were observed. The probe reached the ringed planet in November 1980.

Day after day until early December, images of Saturn flooded into mission control at the Jet Propulsion Laboratory in Pasadena, California. Each day they revealed something new, such as a new moon, ringlets in the ring gaps, and craters on Saturn's moons.

STARPOINT

In September 1997, the Voyagers celebrated their 20th anniversary in space. Voyager 1 was over 6 billion miles (10 billion km) away, and Voyager 2 was nearly 5 billion miles (8 billion km) away.

ANOTHER VOYAGE

By mid-December 1980, Voyager 1's main mission of investigating Saturn was over, and it headed off into outer space. Meanwhile, Voyager 2 was still a long way from Saturn. It had taken a different path through space toward its target. This was planned so that its course would take it not only to Saturn, but on to more distant planets—Uranus and Neptune. So Voyager 2 did not close in on Saturn until the summer of 1981. It repeated Voyager 1's success and returned floods of data and tens of thousands of exciting images.

STARPOINT

NASA had aimed Voyager 2 with great precision. When it made its closest approach to Saturn, on August 25, 1981, it was less than 30 miles (48 km) off target and just 3 seconds ahead of schedule. This was amazingly accurate after a four-year journey of over 1 billion miles (1.6 billion km).

Pictures taken by the Voyagers are stunningly beautiful. This one clearly shows the bands in Saturn's atmosphere.

Inset: This close-up view of Saturn shows the complicated patterns of wind flow in the planet's atmosphere.

Uranus, Neptune, and Pluto

Uranus, Neptune, and Pluto are distant and mysterious worlds. All three lie many millions of miles away from Earth in the outer reaches of the solar system—the family of bodies that circle in space around the Sun. Of the nine planets in the solar system, Uranus, Neptune, and Pluto are the most distant. In our night sky, Uranus looks like a faint star if viewed with the naked eye. Neptune and Pluto can be spotted only with powerful telescopes.

Scientists have discovered that Uranus and Neptune are giant gas planets, similar to Jupiter and Saturn. Uranus and Neptune are much smaller than the other two gas giants, but they still measure about four times bigger across than Earth. And like Jupiter and Saturn, these planets are made up mainly of gas and liquid, with no solid surface.

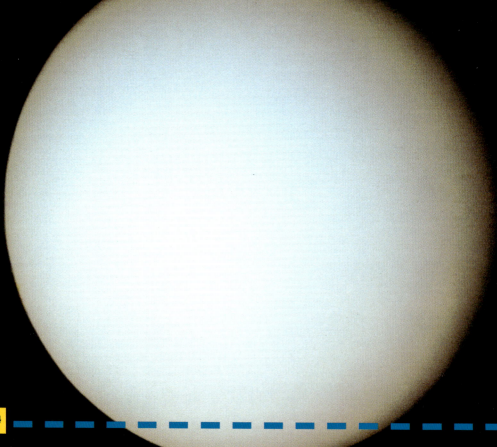

The Voyager 2
space probe took these
pictures of Uranus
(opposite) and Neptune
(right) in 1986 and 1989.

Astronomers knew very little about Uranus and Neptune until the Voyager 2 space probe flew past the planets in the 1980s. The probe discovered systems of rings and many moons circling around both planets.

Pluto is quite different from Uranus and Neptune. Compared to its neighbours, Pluto is tiny. In fact, it has been downgraded from being a planet and is currently being referred to as an asteroid. Its makeup also differs from that of Uranus and Neptune. Unlike these gas giants, Pluto is mixture of ice and rock. Our most distant planet has no rings and just one moon, named Charon. Remarkably, Charon is half the size of Pluto. Astronomers often call Pluto-Charon a double planet.

Discovering New Worlds

Before the discovery of Neptune in 1781, astronomers believed our solar system contained only six planets.

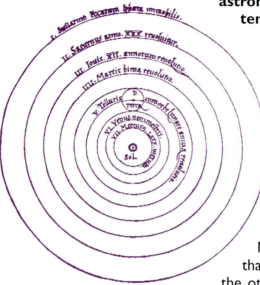

In ancient times, astronomers watched what they thought were exceptionally bright stars wandering across the night sky. They called these wanderers planets. Ancient astronomers knew of five planets—Mercury, Venus, Mars, Jupiter, and Saturn. These planets could be seen with the naked eye.

In the 1500s, Polish astronomer Nicolaus Copernicus discovered that Earth circled the Sun along with the other planets. The solar system thus consisted of six planets. No one suspected that there were other planets waiting to be discovered.

Above: A drawing in one of Copernicus's notebooks shows Earth and five other planets circling around the Sun.

Right: William Herschel, who discovered Uranus in 1781.

PLANET SEVEN

On March 13, 1781, a young musician and astronomer named William Herschel began his evening stargazing at his home in England. That night he spotted an unusual object in the constellation, or pattern of stars, known as Gemini. At first Herschel thought the object was a comet—an icy body that orbits the Sun and blazes in our sky as it approaches the Sun. But soon Herschel and other astronomers realized that he had discovered a seventh planet. It was eventually named Uranus, after the Greek god of the heavens. The planet proved to be twice as far away from the Sun as Saturn, which was then thought to be the most distant planet. So Herschel's discovery that March night had doubled the size of the solar system.

THE SEARCH CONTINUES

For years astronomers tried to determine the exact path Uranus follows around the Sun, but they found that the planet did not move according to their predictions. They came to the conclusion that another planet must be affecting Uranus's orbit. It was possible, they believed, that the gravity of an unknown planet might be tugging Uranus off course. Gravity is the attraction, or pull, one body has on the objects around it.

By the 1840s, two astronomers believed that they had figured out the position of the mystery planet. They were John Couch Adams in England and Urbain Leverrier in France. On September 23, 1846, German astronomer Johann Galle used his telescope to search the part of the sky that Leverrier had suggested. He soon found the mysterious body, planet number eight. It was named Neptune.

THE HUNT FOR PLANET NINE

After Neptune's discovery, astronomers determined that it alone could not account for Uranus's movements. So by the end of the 1800s, the hunt was on for a ninth planet. Among those looking for it was the American astronomer Percival Lowell, who had set up an observatory in Arizona. Lowell had no luck finding the unseen planet before he died in 1916.

Above: In 1845, English mathematician John Couch Adams determined where Neptune could be found.

Frenchman Urbain Leverrier independently calculated Neptune's location at nearly the same time as Adams did.

In 1929, a young astronomer named Clyde Tombaugh began searching for the ninth planet at the Lowell Observatory. His job was to photograph the sky night after night and then examine the photographs to see if any objects had moved among the stars. For many months, Tombaugh took pictures of the region where Lowell had thought the planet might be. He studied these pictures carefully until he found what he was looking for. On February 18, 1930, Tombaugh discovered planet nine, later named Pluto. The observatory announced the discovery of the planet on March 13, 1930—149 years to the day after Herschel had discovered Uranus.

Clyde Tombaugh used a 24-inch telescope at the Lowell Observatory (above) in Arizona to photograph the sky. He discovered Pluto by noticing a body that had changed its position among the stars (below).

A LUCKY CHANCE?

Astronomers soon realized that Pluto's small size and weak gravity hardly affects Uranus's orbit. It was actually by chance that Lowell's calculations placed Pluto in the constellation Gemini, where Tombaugh later found it. For years afterward, astronomers continued to search for a tenth planet that would explain Uranus's movements.

Uranus

Uranus is the third largest planet in the solar system, but its great distance from Earth makes it difficult to see with the naked eye.

Uranus is the seventh planet in the solar system, going out from the Sun. At a distance of about 1.8 billion miles (2.9 billion km), it lies twice as far away from the Sun as Saturn. Uranus takes 84 Earth-years to travel once around the Sun.

direction of orbit

direction of rotation on axis

axis

   If you know exactly where to look, you can just barely see Uranus on a clear night. The planet is more easily seen through binoculars or a small telescope. Even in large telescopes, Uranus only appears as a pale greenish body, with no obvious markings.

Above: Uranus rotates on an axis that is nearly tilted on its side. Sometimes the planet's poles point directly toward the Sun.

TIPPING OVER

Like the other planets, Uranus rotates, or spins around on its axis. A planet's axis is an imaginary line through it from its north pole to its south pole. But Uranus has an unusual way of rotating on its axis. Most planets rotate on an upright or slightly tilted axis as they orbit.

   Uranus's axis is tilted so far over that the planet rotates nearly on its side as it travels around the Sun. This means that the poles on Uranus face either directly toward or away from the Sun. When one pole faces the Sun the other faces away.

Neptune

Uranus

Saturn

Right: Uranus lies in the solar system between Saturn and Neptune. It is twice as far from the Sun as Saturn is and takes nearly three times as long to complete its orbit.

Right: Uranus is about four times bigger across than the Earth. It has more than 15 times the Earth's mass.

SMALL GIANT

Uranus is the third largest planet in the solar system. With a diameter, or distance across, of about 31,700 miles (51,100 km), it is one of the four giant planets. In fact, Uranus could swallow over 60 bodies the size of Earth. This makes the planet slightly bigger than Neptune but much smaller than the two largest planets, Jupiter and Saturn.

INSIDE URANUS

Like Jupiter and Saturn, Uranus is surrounded by a thick atmosphere, or layer of gases. The planet's atmosphere contains mostly hydrogen and helium. A small amount of methane gas in the atmosphere gives Uranus its blue-green appearance.

Astronomers do not know exactly what lies beneath Uranus's atmosphere. But its surface probably differs from that of Jupiter or Saturn. These two giant planets most likely have deep oceans of liquid hydrogen. In contrast, Uranus may have an ocean made up of water, methane, and ammonia. At the centre of the planet lies a core of liquid rock about the size of Earth.

Beneath Uranus's thick atmosphere is probably a deep ocean of water and icy gases. The liquid-rock core may contain some metal.

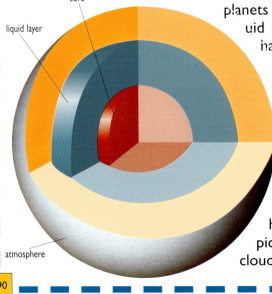

core

liquid layer

atmosphere

THE WEATHER

Compared with Jupiter and Saturn, Uranus has mild weather. Close-up pictures of the planet reveal few weather features in its hazy atmosphere. However, some pictures have shown faint bands of clouds deeper in the atmosphere. These

Magnetic Uranus

All around Earth is an invisible force called magnetism, which is created by our planet's iron core. The magnetism extends into outer space to form a magnetic bubble around Earth known as the magnetosphere. Uranus has a magnetosphere, too. But unlike the magnetosphere on Earth and many other planets, Uranus's magnetosphere does not line up with its north and south poles. Instead, its magnetic poles are closer to the planet's equator. Astronomers have not discovered exactly what produces Uranus's unusual magnetism.

clouds rotate in the same direction as the planet. They are driven by winds blowing at speeds of up to about 370 miles (600 km) per hour. But pictures of Uranus have not revealed anything like the storms that rage on Saturn and Jupiter.

In ordinary photographs (left), Uranus appears to be the same colour all over. But a few clouds show up in computer-processed photographs (above).

AN ACCIDENTAL DISCOVERY

On March 10, 1977, American astronomers were flying over the Indian Ocean in a plane carrying a powerful telescope. They planned to observe Uranus as it passed in front of a particular star. By measuring the time Uranus took to pass in front of the star, they could determine the planet's size.

The astronomers switched on their measuring instruments before Uranus was due to pass in front of the star and block out its light. Then a strange thing happened. The star seemed to wink, or dim, several times and then become bright again. as if something were blocking the star's light. After Uranus had passed the star, the star appeared to wink several times again. Astronomers eventually discovered that the winks were caused by a set of rings. Until that time, Saturn was the only planet known to have rings.

An artist's impression of the rings around Uranus. The rings are actually much less bright than they appear here. They are made up of dark material that does not reflect sunlight well.

URANUS'S RINGS

Astronomers have found 11 rings circling around Uranus. The rings begin about 7,000 miles (11,000 km) above Uranus's cloud tops. Altogether, the ring system is about 8,000 miles (13,000 km) wide. The faint innermost ring appears to be made up of fine dust. This ring is about 1,500 miles (2,500 km) wide. The other rings are much narrower, mostly a few miles wide. They are made up mainly of dark boulders that measure a few feet across.

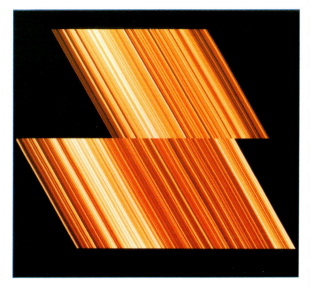

Above: These colourized pictures show slices through Uranus's outermost ring. Computer processing has revealed many ringlets within the ring.

Left: Wide sheets of fine dust appear between Uranus's narrow rings when they are lit from behind by the Sun. The short streaks of light are trails made by distant stars.

Above: The orbits of Uranus's five largest moons. Miranda, the closest of the moons, takes less than a day and a half to orbit Uranus; Oberon takes more than 13 days.

URANUS'S MOONS

Scientists have discovered 17 moons orbiting Uranus. Five of them are fairly large and were observed by early astronomers. William Herschel, Uranus's discoverer, discovered the planet's two largest moons in 1787. They were named Titania and Oberon. Two more moons, Ariel and Umbriel, were discovered in 1851, and a fifth moon, Miranda, was spotted in 1948. Uranus's other twelve moons are too small to be seen from Earth. Even the largest, Puck, is only about 95 miles (150 km) across. And the smallest, Cordelia, is only about 15 miles (25 km) across.

Uranus's five largest Moons are small compared with Earth's Moon.

AMAZING MIRANDA
Miranda is one of the strangest moons in the entire solar system. Its surface is a patchwork of many different kinds of landscape. Some parts of its surface are covered in craters, or deep pits, much like the surface of the Moon. Other parts are rugged, with steep cliffs and deep cracks in the surface. Perhaps the most unusual feature on Miranda is a set of curving grooves that looks like a giant oval race-track on the moon's surface. Some of the oval grooves measure up to 200 miles (320 km) across.

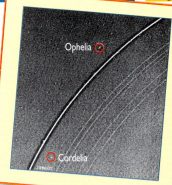

The Little Shepherds

Two of Uranus's tiny moons, Cordelia and Ophelia, orbit close to Uranus's outermost ring. They are called shepherd moons because of the way they seem to herd the ring particles. As these moons travel around either side of the ring, their gravity helps pull back any particles that stray outside it.

STARPOINT

Most of Uranus's moons are named after characters from the plays of the famous British playwright William Shakespeare.

This is how Uranus might look from Miranda. An artist has added Uranus's rings, although they would probably be too faint to be seen from Miranda.

ARIEL

Ariel is the brightest of Uranus's moons. Sheets of water ice have created bright streaks and patches on its surface. Like the surface of Uranus's other large moons, Ariel's surface is covered with craters. Most craters on this moon are less than 20 miles (30 km) in diameter. Deep cracks and a number of wide, branching valleys with smooth floors also cut into Ariel's surface.

UMBRIEL

Umbriel is the same size as Ariel, but it has a much darker surface. In fact, it is the darkest of Uranus's moons. Much of Umbriel's surface is covered with large

After Miranda, Ariel has the most interesting surface among Uranus's moons. The two main valleys seen here are Korrigan Chasma (left) and Kewpie Chasma.

MOONS OF URANUS DATA				
Moon	Diameter (miles)	(km)	Ave. distance from planet (miles)	(km)
Miranda	300	480	80,400	129,400
Ariel	720	1,160	118,700	191,000
Umbriel	730	1,170	165,200	266,000
Titania	980	1,580	271,000	436,300
Oberon	950	1,520	362,300	583,400

Umbriel

Titania

Oberon

craters such as Skynd, which is at least 70 miles (110 km) across. The strangest feature on the moon is a bright circle called the "fluorescent Cheerio." Astronomers think that this feature may be the icy rim or floor of a crater.

TITANIA

Titania is Uranus's largest moon. In many ways, it looks similar to Ariel because deep cracks and valleys mark its surface. The largest valley, named Messina Chasma, stretches for more than 900 miles (1,400 km). That's over three times the length of Earth's Grand Canyon in Arizona. Some of Titania's valleys cut through the many craters that cover the moon. The largest craters are up to 185 miles (300 km) wide.

OBERON

Of the large moons, Oberon orbits farthest from Uranus. Deep cracks mark Oberon's heavily cratered surface, and an unidentified dark coating covers the floors of some craters on the moon. The coating may be a mixture of ice and materials containing carbon.

Neptune

The most distant of the gas giants, Neptune is very similar to Uranus in size and makeup.

Neptune is the eighth planet in the solar system. It lies on average about 2.8 billion miles (4.5 billion km) from the Sun and takes nearly 164 Earth-years to complete its orbit. Neptune's day, or the time it takes to rotate once on its axis, is short—just over 16 Earth-hours.

Only slightly smaller than Uranus, Neptune measures about 30,800 miles (49,500 km) across. But this gas giant is too far from Earth to be seen with the naked eye. Even with a powerful telescope, we can see few details on the planet. We have learned what we know about Neptune from the Voyager 2 space probe that travelled to the outer reaches of the solar system.

Like Uranus, Neptune is about four times bigger across than Earth.

In the solar system, Neptune lies between Uranus and Pluto. At certain times, Pluto journeys inside Neptune's orbit, and Neptune becomes the farthest planet.

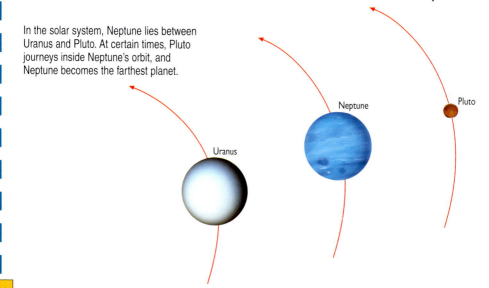

Uranus

Neptune

Pluto

Neptune's Makeup

The blue face of Neptune that we see in photographs is the top of a very thick atmosphere. Similar to the atmosphere of Uranus, Neptune's atmosphere contains mainly hydrogen and helium, with a smaller percentage of methane. As on Uranus, the methane gas gives Neptune its lovely blue colour.

A deep ocean of water, methane, and ammonia probably covers Neptune beneath its thick atmosphere. Underneath this ocean lies an Earth-sized core made of liquid rock and ice.

Heated Planet

By our standards, Neptune is a very chilly place. But its temperatures are higher than we would expect for a planet so far from the Sun. That's because Neptune appears to be heated from within. The planet releases more than twice as much heat as it receives from the Sun. At its cloud tops it temperature measures about −350° F (−210° C). This is very similar to temperatures on Uranus, even though Uranus is a billion miles (1.6 billion km) closer to the Sun.

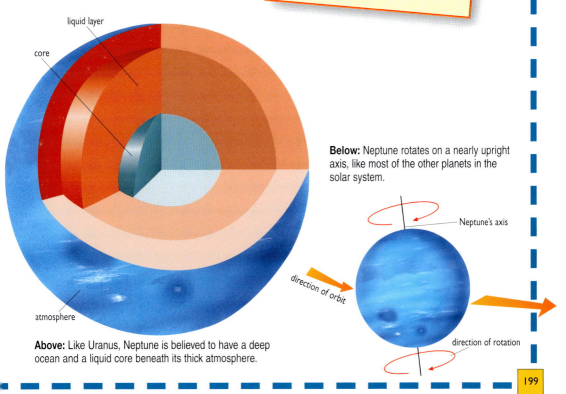

liquid layer

core

atmosphere

Below: Neptune rotates on a nearly upright axis, like most of the other planets in the solar system.

Neptune's axis

direction of orbit

direction of rotation

Above: Like Uranus, Neptune is believed to have a deep ocean and a liquid core beneath its thick atmosphere.

Above: An unusual view of the clouds in Neptune's atmosphere, with the Sun shining on them from a low angle. The shadows they cast make them stand out clearly.

WIND AND CLOUDS

Heat produced within Neptune creates violent winds in its atmosphere. They whip around the planet in the opposite direction of Neptune's rotation. The winds can reach speeds of 1,500 miles (2,400 km) per hour, making them the fastest winds in the solar system.

Along with strong winds, bands of high white clouds rotate in Neptune's atmosphere. The clouds are made up of methane ice, in much the same way that high cirrus clouds on Earth are made up of water ice. One cloud patch, named Scooter, flies around the planet once every 16 hours. It appears to change size and shape over time.

STORMY WEATHER

Furious storms break out in Neptune's windy atmosphere. The storms appear on Neptune as dark oval regions surrounded by wisps of white cloud. In 1989, scientists observed one particularly large storm in the planet's atmosphere. They named it the Great Dark Spot (GDS) after the great storm on Jupiter called the Great Red Spot.

Right: A close-up picture of Neptune's Great Dark Spot, a huge storm system in the planet's atmosphere. It is ringed with clouds.

The GDS was not as big as the Great Red Spot, but it was still very large—about the size of Earth. Winds raced around the GDS at speeds of up to 745 miles (1,200 km) per hour. While Jupiter's Great Red Spot has lasted for hundreds of years, storms on Neptune appear to have short lives. Pictures taken by the Hubble Space Telescope showed that the GDS had disappeared by 1994.

ANOTHER RINGED PLANET?

By the 1980s, scientists had discovered that Saturn, Jupiter, and Uranus all had rings. Astronomers began wondering whether Neptune had rings, too. From Earth, no rings could be seen around the planet.

Astronomers searched for rings around Neptune using the same method that had led them to discover Uranus's rings. They waited for a star to wink just before and just after Neptune passed in front of it. The results of this experiment were confusing. When Neptune passed in front of a star, scientists observed that the star winked sometimes but not all the time. They concluded that instead of having complete rings, Neptune must have arcs, or part-rings.

NEPTUNE DATA

Diameter at equator:
30,800 miles (49,500 km)
Average distance from Sun: 2,794,000,000 miles (4,498,000,000 km)
Rotates on axis in:
16 hours, 7 minutes
Orbits Sun in: 163.7 years
Moons: 8

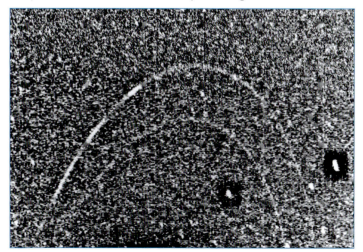

Voyager 2's first clear pictures of Neptune's rings show clumps of bright material that appeared at first to be arcs around the planet.

VOYAGER'S VIEW

When the space probe Voyager 2 approached Neptune in 1989, it spotted the ring arcs astronomers had detected from Earth. As the probe got closer, astronomers could see that the arcs were actually part of a complete ring. Voyager 2 discovered that Neptune had four rings in all.

The two main rings are bright and narrow and appear to be made up of fine dust and small particles. The outer ring contains bright clumps of particles. These clumps are what astronomers mistook for the arcs. Neptune's other two rings are wide but much fainter. The one closest to the planet may be at least 1,100 miles (1,700 km) wide. The other faint ring, between the two narrow main rings, measures about 3,600 miles (5,800 km) wide.

Above: This Voyager picture shows Neptune's two main rings. The outer ring lies about 39,000 miles (63,000 km) from the planet's centre, while the other ring is about 33,000 miles (53,000 km) closer.

NEPTUNE'S MOONS

From telescopes on Earth, only two moons can be seen orbiting Neptune—Triton and Nereid. Triton, Neptune's largest moon, is about two-thirds the size of our Moon. It measures 1,680 miles (2,700 km) across. Nereid is only about 210 miles (340 km) across. Neptune's six remaining moons were discovered by

Below: Of the four rings that circle Neptune, two are narrow and bright and two are broad and faint.

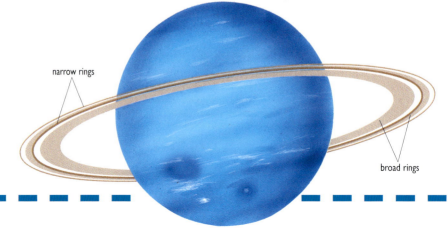

narrow rings

broad rings

the Voyager 2 space probe. They are Naiad, Thalassa, Despina, Galatea, Larissa, and Proteus.

At 260 miles (420 km) across, Proteus is actually larger than Nereid. It cannot be spotted from Earth because it lies too close to Neptune and gets lost in the planet's glare. The rest of the moons are very small—from 30 to 240 miles (60–180 km) across. These small moons are made up of a dark, rocky material, and they have an irregular shape.

DEEP-FROZEN TRITON

Triton orbits Neptune at a distance of about 220,000 miles (354,000 km) from the planet, about the same distance between the Moon and Earth. However, while the Moon orbits Earth in the same direction as our planet's rotation, Triton orbits in the opposite direction of Neptune's rotation.

Triton is a deep-frozen world with a temperature of only about −390° F (−235° C). This is the coldest known place in the solar system, colder even than distant Pluto. A layer of frozen gases, including nitrogen and methane, covers its icy surface. A pink-coloured polar cap rises at the moon's south pole. And icy volcanoes cover parts of the moon where gases, ice, and plumes of dust forced their way to the surface. Material from the plumes settled on the ground and created dark streaks on the surface. Deep cracks, some of them filled with ice, also cut into Triton's frozen landscape.

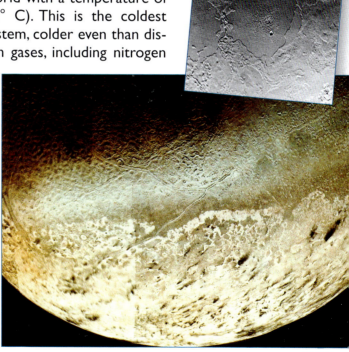

Two views of Neptune's largest moon, Triton. The bottom picture shows the region around the moon's south pole. The inset shows a huge ice-filled crater.

Pluto

Astronomers know very little about tiny Pluto, the most distant of the planets. Some astronomers wonder if it should even be classified as a planet.

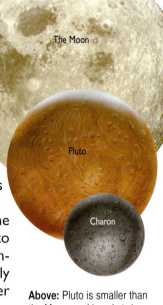

The Moon

Pluto

Charon

Above: Pluto is smaller than the Moon, and is only twice as big as its own moon, Charon.

Pluto is by far the smallest planet in the solar system. With a distance across of only about 1,480 miles (2,390 km), it's less than half the size of the next largest planet, Mercury, and only two-thirds the size of our Moon.

Most of the time, Pluto is the outermost planet in the solar system. However, its unusual orbit causes it to occasionally change positions with its closest neighbour, Neptune. Pluto orbits the Sun in an extremely elliptical, or oval, orbit—more elliptical than any other planet in the solar system. This means that its distance from the Sun varies greatly, from a near point of about 2.8 billion miles (4.5 billion km) to a far point of about 4.6 billion miles (7.4 billion km). At times, Pluto's oval orbit takes it inside the orbit of Neptune. When this occurs, Neptune was the outermost planet.

Pluto's orbit differs from that of the other planets in another way. It travels in a different plane. A plane is like an imaginary sheet of paper in space, with the Sun in the centre of the paper and the planets orbiting around it. Most of the planets orbit near to this same imaginary piece of paper, or the same plane. Pluto orbits in a different plane. Its orbit takes it far above and far

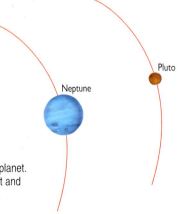

Pluto

Neptune

Right and opposite top: Most of the time, Pluto is the farthest planet. But in February 1979, Neptune journeyed outside of Pluto's orbit and became the outermost planet, which lasted until February 1999.

Pluto's orbit

Neptune's orbit

1979 crossing

Sun

1980

1999 crossing

2000

below the other planets during its 248-year journey around the Sun.

Some scientists wonder if Pluto is a planet at all. In the past, astronomers had suggested that Pluto was a moon that escaped from Neptune's orbit. More recently, astronomers have thought that Pluto may be one of the leftover bodies from the solar system's formation. Nearly 5 billion years ago, the solar system formed from a great cloud of gas and dust. Most of this material formed the Sun, the planets, and their moons. The leftover smaller lumps of matter are known as asteroids, comets, and meteors. It is possible that Pluto may be the one of largest of these leftover lumps of matter. However, until very recent times most scientists have classified Pluto as a planet because of its size, its predictable orbit, and the fact that it has a moon.

The powerful Hubble Space Telescope is the only instrument that can picture Pluto (left) and its moon, Charon, clearly.

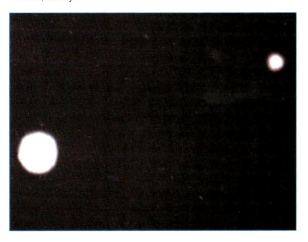

LOOKING AT PLUTO

Even in the most powerful telescopes on Earth, distant Pluto looks like a faint star. No details of the planet's surface can be observed from Earth. However, the Hubble Space Telescope, orbiting above Earth's atmosphere, has taken pictures that reveal some surface details. These pictures have shown that Pluto is an icy planet with polar caps. It also has light and dark areas on its surface, probably caused by patterns of ice and frost.

PLUTO'S MAKEUP

Pluto is a dark, cold world made up of a mixture of rock and ice. Temperatures on the planet are around −355° F (−215° C). Frozen methane, nitrogen, and carbon monoxide gas probably make up Pluto's outer layer of ice. Beneath this outer layer lies what may be a layer of water ice. Underneath that may be a rocky core.

Pluto has a very faint atmosphere made of nitrogen, carbon monoxide, and methane—the same gases that make up Pluto's top layer of ice. Astronomers believe that Pluto's atmosphere forms as it approaches the Sun and disappears as it travels away from the Sun. When Pluto is closest to the Sun, the Sun's heat melts some of the ice on Pluto's surface. The melted ice changes into gases that rise from the surface to form the thin atmosphere. As Pluto travels away from the Sun, the gases freeze again and fall back to the surface.

core

frozen methane, nitrogen, and carbon monoxide

frozen water ice

Above: Astronomers think that Pluto is probably made up of several layers of ice surrounding a rocky core.

A DOUBLE PLANET?

In 1978, astronomers studying magnified pictures of Pluto noticed that the planet had an odd shape. Pluto appeared to have a bump on one side. Clearer pictures

Left: In telescopes on Earth, the images of Pluto and Charon look blurred together (top left). But the Hubble Space Telescope shows that the two bodies are separate.

revealed that the bump was not attached to Pluto. It was actually a moon, which astronomers called Charon.

Remarkably, Charon turned out to be half the size of Pluto. No other planet has a moon that is half its size. Until astronomers discovered Charon, our Moon was thought to be the largest moon in comparison with its planet—and it is only one-fourth of Earth's size. Charon is so big compared to Pluto that some astronomers call the two bodies a double planet.

Charon orbits very close to Pluto, only about 12,200 miles (19,600 km) away. It completes an orbit around Pluto in about 6 days and 9 hours. Amazingly, Pluto takes only slightly more time to rotate once on its axis. This means that the planet and moon appear to be locked together as they move through space. Charon remains in nearly the same location above Pluto.

PLUTO DATA

Diameter at equator:
1,484 miles (2,390 km)
Average distance from Sun:
3,700,000,000 miles
(5,900,000,000 km)
Rotates on axis in:
6 days, 9 hours, 36 minutes
Orbits Sun in: 248 years
Moons: 1

An artist has painted this picture of Pluto and Charon, billions of miles away, the Sun shows up as a bright star.

STARPOINT

If you could visit Pluto, you would only be able to see Charon from one side of the planet. Charon would never be visible from the other side.

Voyager's Distant Mission

Voyager 2 has provided us with most of our information about distant Uranus and Neptune.

In 1977, NASA began its missions to the outer planets with the launch of Voyager 2 in August and Voyager 1 in September. Most of the time, the outer planets are too far apart for a space probe to visit all of them in one mission. But the launchings were timed to take advantage of the outer planets' positions in space. Voyager 1 visited Jupiter and Saturn, while Voyager 2 was chosen to travel to all four gas giants. Unfortunately, the probe was not able to visit Pluto.

Voyager 2 closes in on Uranus in January 1986. It is heading towards the planet's south pole, which was facing the Sun at the time. It is now so far away that its radio signals, travelling at the speed of light, take 2 hours and 45 minutes to travel back to E

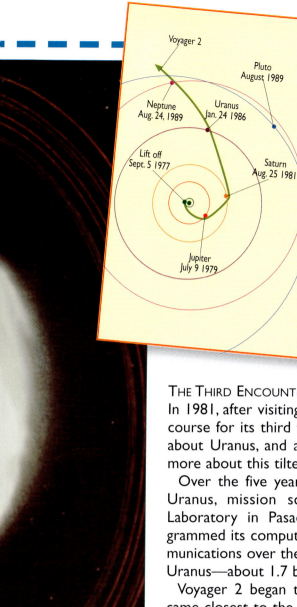

Voyager 2

Pluto
August 1989

Neptune
Aug. 24, 1989

Uranus
Jan. 24 1986

Lift off
Sept. 5 1977

Saturn
Aug. 25 1981

Jupiter
July 9 1979

Gravity Assist

The method NASA used to get Voyager 2 from planet to planet is called gravity assist, or the slingshot method. Scientists planned for the space probe to use the gravity of planets. While passing near a planet, the probe would be speeded up and flung away by the force of the planet's gravity. Scientists were so successful with this method that Voyager 2 reached its final target, Neptune, within a few seconds of the planned schedule. By then, the probe had travelled for more than 12 years over a distance of more than 4 billion miles (7 billion km)!

THE THIRD ENCOUNTER

In 1981, after visiting Jupiter and Saturn, Voyager 2 set course for its third target—Uranus. Little was known about Uranus, and astronomers were eager to learn more about this tilted planet.

Over the five years it took Voyager 2 to travel to Uranus, mission scientists at the Jet Propulsion Laboratory in Pasadena, California, carefully reprogrammed its computer. They wanted to improve communications over the vast distance between Earth and Uranus—about 1.7 billion miles (2.7 billion km).

Voyager 2 began to approach Uranus in 1985 and came closest to the planet in January 1986. The space probe revealed that the planet was an overall blue-green colour, with no obvious features. Voyager 2 also discovered more rings and 10 tiny moons. Its pictures of the moon Miranda caused the greatest surprise. Miranda proved to have one of the strangest surfaces of all the moons in the solar system.

Asteroids, Comets, and Meteors

Nearly 5 billion years ago, the solar system began to form from a great cloud of gas and dust. Most of this material formed the Sun, the star at the centre of the solar system. The rest of the material continued circling the Sun, eventually forming the nine planets and their moons. After these larger bodies formed, many smaller lumps of matter were left over. These lumps are what we call asteroids, comets, and meteors. They were scattered throughout space between the planets and at the edge of the solar system. Many of the smaller leftover particles rained down on the newly formed planets and moons.

Asteroids, comets, and meteors are made up mostly of rock, ice, and metal. Asteroids and meteors are mostly rock, but some also contain metal. Comets are a mixture of rock, ice, and dust. Scientists sometimes call them "dirty snowballs." Scientists study the ancient materials in these small worlds to learn more about how our solar system began.

Above: The asteroid Dactyl, which was discovered by the space probe Galileo on its way to Jupiter, measures about 1 mile (1.6 km) across.

Sun

Earth

Mars

Jupiter

asteroid belt

Most asteroids are found in a broad ring, or belt, between the orbits of Mars and Jupiter.

The Asteroids

Most asteroids circle the Sun in a wide band between the orbits of Mars and Jupiter. Scientists have identified more than 7,000 of these rocky bodies, but as many as a million more may exist.

The asteroids are the largest of the small worlds that orbit the Sun. But even Ceres, the biggest asteroid, is less than one-third the size of the Moon and about half the size of the smallest planet, Pluto. Most asteroids, however, are much smaller than Ceres. Many have a diameter, or distance across, of a few tens of miles.

DISCOVERING ASTEROIDS
In 1801, an Italian astronomer named Giuseppe Piazzi discovered the first asteroid. An astronomer is a scientist who studies outer space. Piazzi thought the object was a comet, but it proved to be a small body orbiting

Below: Even the three biggest asteroids—Ceres, Pallas, and Vesta—are very small compared to Earth.

Pallas

Ceres

Vesta

about midway between the orbits of Mars and Jupiter. He named it Ceres. Then, in 1802, German astronomer Heinrich W. M. Olbers noticed another bright object also between the orbits of Mars and Jupiter. Like Piazzi, he first believed the object was a comet, but it turned out to be another kind of small body. Olbers named the orbiting object Pallas. He used the word asteroid, which means "starlike," to describe his discovery. Over the next four years, two more asteroids, Juno and Vesta, were discovered orbiting between Jupiter and Mars. Since that time, thousands of asteroids have been found.

The total weight of all the asteroids put together is less than the weight of the Moon.

THE BIG THREE

Scientists have discovered at least 26 asteroids that are larger than 125 miles (200 km) in diameter. The rest range in size from less than 125 miles to just over ½ mile (1 km) across. The largest three asteroids are Ceres, Pallas, and Vesta. Ceres, by far the largest, has a diameter of about 600 miles (950 km). Pallas and Vesta are both about 340 miles (550 km) in diameter. Scientists believe that these three large asteroids are nearly ball shaped. But most asteroids have an irregular shape, sort of like a potato. Others appear to be longer and narrower.

Left: Giuseppe Piazzi (1746-1826) discoved the first asteroid, Ceres, in 1801.

ASTEROIDS DARK AND BRIGHT

Asteroids are divided into three main groups according to their makeup. Most asteroids are dark, stony rocks called C-types. These asteroids are difficult to see because they are dull in colour and do not reflect much of the Sun's light. A second group, the S-type asteroids, are brighter than the C-types and tend to be easier to see. They are gray in colour and contain some metal. The rarest asteroids—the M-types—are also the brightest. They are made of pure nickel-iron metal.

Below: This colour map of the asteroid Vesta shows the elevation, or height, of different parts of its surface. The bright colours represent the highest regions.

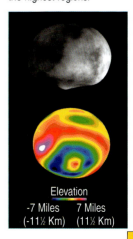

Elevation

-7 Miles   7 Miles
(-11½ Km)   (11½ Km)

Amor
main asteroid belt
Eros
Ceres
Adonis
Apollo
Hidalgo

Above: Most asteroids, including the largest, Ceres, orbit within the asteroid belt. But some asteroids have orbits that take them inside or outside the belt.

ASTEROID ORBITS

Most asteroids orbit the Sun between Mars and Jupiter in an area called the asteroid belt. The middle of the belt lies about 250,000,000 miles (400,000,000 km) from the Sun. Occasionally, asteroids are knocked out of the asteroid belt when they collide with one another. Others may be tugged off course by the gravity, or pull, of a larger body, such as Jupiter. These asteroids are tugged away from the asteroid belt into the outer solar system or pulled closer to Earth into the inner solar system.

A number of the asteroids orbiting in the inner solar system cross paths with Earth's orbit. Astronomers call these Near-Earth asteroids. A few have come closer than 100,000 miles (160,000 km) to Earth. This puts them dangerously near our planet. Although the chances are quite small, astronomers believe that an asteroid could collide with Earth and in the future cause enormous damage.

THE TROJANS

asteroid belt
Trojans
Mars
Sun
Jupiter
Trojans
orbit of Jupiter

In addition to the asteroids in the asteroid belt and those that cross Earth's orbit, there are two small groups of asteroids that travel in the same orbit as Jupiter. They are known as the Trojans. One group travels in front of Jupiter, the other travels behind the planet. Astronomers

The two groups of Trojan asteroids have been captured by Jupiter's powerful gravity and circle the Sun in Jupiter's orbit.

know of several hundred Trojan asteroids, but they believe that 1,000 or more may exist.

MOONS OR ASTEROIDS?

Most asteroids orbit the Sun, but some asteroids may orbit certain planets as moons. Some astronomers think that Mars's two small moons, Phobos and Deimos, were once asteroids from the nearby asteroid belt. They might have been captured by the pull of Mars's gravity when they came close to the planet billions of years ago. Some of the small outer moons of Jupiter, Saturn, Uranus, and Neptune may once have been asteroids, too.

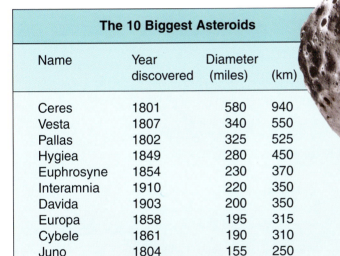

The 10 Biggest Asteroids			
Name	Year discovered	Diameter (miles)	(km)
Ceres	1801	580	940
Vesta	1807	340	550
Pallas	1802	325	525
Hygiea	1849	280	450
Euphrosyne	1854	230	370
Interamnia	1910	220	350
Davida	1903	200	350
Europa	1858	195	315
Cybele	1861	190	310
Juno	1804	155	250

Phobos, the largest of Mars's two moons, is probably a captured asteroid. It measures less than 20 miles (30 km) across.

Our Moon

Some scientists believe that an asteroid may have created our Moon. Billions of years ago, a very large asteroid may have crashed into Earth and knocked off large rocky pieces of the planet. These rocky chunks were flung into space. In time, they may have come together to form a separate body, the Moon.

Left: The leftover lumps of rock and metal between Mars and Jupiter never formed into another planet. Instead, they continued to orbit the Sun as asteroids.

WHERE ASTEROIDS CAME FROM

For many years, astronomers thought that the asteroids were the remains of another planet. This planet began forming between Mars and Jupiter, they said, at the same time as the other planets were forming. Then it gradually moved closer and closer to Jupiter. Jupiter's very strong gravity put a great deal of pressure on this planet. The gravity created powerful tides inside the planet, which eventually pulled it apart into pieces. Scientists believed that these pieces became the asteroids.

Most astronomers no longer believe that the asteroids are pieces from an exploded planet. Instead, they think that the asteroids are leftover lumps of matter in the solar system that never formed into one larger body.

Life of an asteroid:
1. Small pieces of rock and metal fuse together to form an asteroid.
2. The asteroid melts, and any metal it contains sinks to the centre.
3. A cross-section of the newly formed asteroid shows the solid iron core and the rocky outer layer.

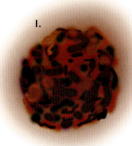

1.

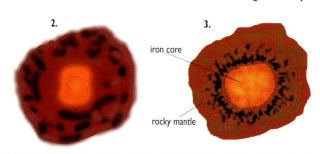

2.

3.

iron core

rocky mantle

In the early days of the solar system, a disk of matter containing gas, dust, and small lumps of rock and metal circled the newborn Sun. Within this disk, the lumps kept colliding and sticking together, gradually forming larger and larger bodies. In time, these bodies formed into the planets. But the lumps of matter between Mars and Jupiter never came together as a planet. When these lumps of rock and metal collided together, Jupiter's powerful gravity pulled them apart.

Above: This distant moon of Saturn is probably a captured asteroid. Its craters show that it has been battered repeatedly in collisions with other asteroids.

LIFE STORY

These newly formed asteroids were made up of rock, dust, and metal bound together. Over time, some of the bigger asteroids probably melted. At this stage, the heavy metal in the large asteroids sank to the centre to form a metal core. The lighter rock formed layers around the core, and the whole asteroid slowly cooled and became solid.

But few asteroids have remained as they were during their early formation. Collisions among asteroids broke them up into smaller pieces and sometimes shattered them completely. In some cases, when a big asteroid shattered from a powerful collision, its metal core was exposed, creating a rare M-type asteroid. In other cases, the rocky crust was not entirely broken away after an asteroid collision, resulting in the more common rocky types of asteroids.

The most violent collisions between asteroids can smash them to pieces and produce hundreds of smaller fragments.

Comets

Comets are a mixture of rock and dust bound together with ice and frozen gases. They visit our skies regularly, sometimes shining more brightly than the stars.

Comets can be a spectacular sight. At their brightest, they are easily seen with the naked eye for weeks or months at a time. In ancient times, the sudden appearance of a comet terrified many people. They believed a comet was a bad sign and that wars, disease, and other disasters would follow.

It is easy to see why comets impressed ancient peoples. Comets can look extremely large—their tails can appear to stretch halfway across the sky. A comet's brightest parts blot out the light of hundreds of stars. However, comets appear to be much bigger than they are. A comet's solid core is often only several tens of miles in diameter, smaller than most bodies in the solar system.

Inside a comet's bright coma, or head, is a tiny nucleus. Well-developed comets have two tails. The bright one is made up of dust, while the faint one consists of glowing gas.

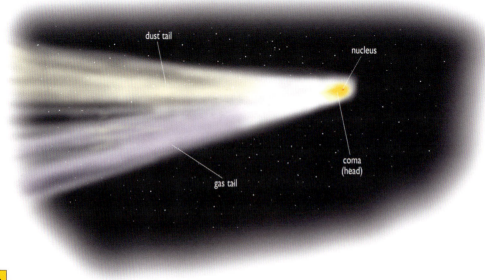

dust tail

nucleus

coma (head)

gas tail

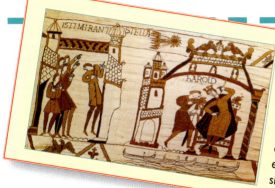

Harold's Comet

Ancient Britons believed they had good reason to fear the comet that appeared in 1066. In that same year, Britain was invaded by the French. The French killed the British ruler, King Harold, and conquered the Britons. The comet appears in the Bayeux Tapestry, a large embroidered cloth that illustrates the invasion of ancient Britain.

LOOKING AT COMETS

A comet has three main parts—the nucleus, the head, and the tail. The two visible parts are the head and the tail. The head, which astronomers call the coma, is the brightest part. It can measure tens of thousands of miles across. Deep in the core of the coma is the only solid part of the comet, the nucleus. The nucleus is small, sometimes measuring less than 10 miles (16 km) across.

Streaming away from the head is the comet's tail, which can grow to enormous lengths of millions of miles. Many comets grow two tails, one straight and the other curved.

Bennett's Comet was the brightest comet of 1970. At the time this picture was taken, only the dust tail was clearly visible.

long-period comet

orbit of comet
Stephan-Oterma

Sun

orbit of comet
Encke

orbit of Saturn

orbit of Uranus

orbit of Neptune

orbit of Halley's comet

orbit of Pluto

ORBITING THE SUN

Like the planets, comets travel through space in an orbit that takes them around the Sun. Comets that take less than 200 years to complete their orbit are called short-period comets. They are regular visitors to our skies. The best known short-period comet is Halley's comet, which returns to our skies about every 76 years.

Other comets follow extremely long orbits. Their journeys around the Sun can take thousands or even millions of years. A comet that takes more than 200 years to orbit the Sun is called a long-period comet. Long-period comets travel to distant reaches of the solar system before heading back in toward the Sun.

STARTING TO SHINE

During most of its orbit, a comet is invisible to us. It has no coma or tail, and its small nucleus cannot be seen from Earth, even with powerful telescopes.

STARPOINT

Encke's Comet has the shortest known orbit of any comet. It orbits the Sun in just 3.3 years.

Comet Kohoutek

In 1973, a long-period comet named Kohoutek was first spotted in telescopes when it was more than 430 million miles (700 million km) from the Sun. Comets cannot usually be detected at such a distance, so astronomers predicted that it was going to be the brightest comet of the century. But when the comet did approach the Sun, scientists were surprised to find that it was barely visible to the naked eye. Astronauts in the U.S. space station Skylab studied this comet, making it the first comet to be studied closely from space. Kohoutek will not return to Earth's skies for at least another 75,000 years.

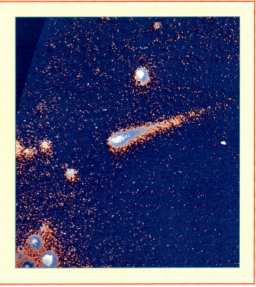

beginning of cloud

cloud expands

A cloud of gas and dust gradually grows around the nucleus of a comet as it approaches the Sun. As the cloud reflects sunlight, it becomes visible in our skies.

comet travels towards the Sun

But as an icy comet gets within a few hundred million miles of the Sun, things start to change. The Sun's heat warms the outer surface of the comet and melts some of its ice. The ice changes into gas and forms a cloud around the nucleus. At the same time, the dust that was frozen in the ice is released in the gas. This cloud of gas and dust begins to reflect sunlight. As the comet comes closer to the Sun, the cloud around the nucleus grows larger and shines more brightly in the sunlight.

cloud starts to be pushed away from nucleus

sunlight

The tail of a comet is longest when it is closest to the Sun, and it always points away from the Sun.

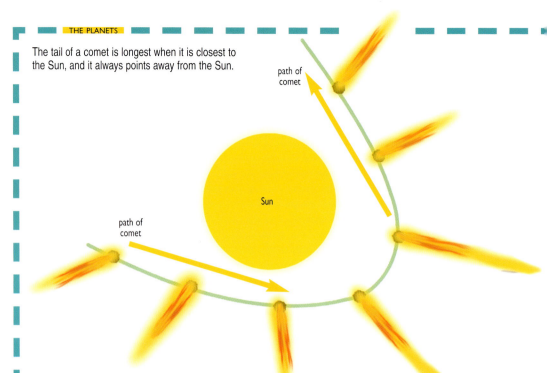

path of comet

path of comet

Sun

SUN PRESSURE

The Sun also has another effect on the cloud of gas and dust around the comet. Pressure from the Sun's radiation and the solar wind (a fast-moving stream of particles that move outward from the Sun) gently press against the cloud. These forces push some of the cloud's dust and gas away from the main body of the comet. Over time, a stream of dust and gas forms the comet's tail.

The Sun's radiation and the solar wind push against the particles of gas and dust around a comet's nucleus, forming the comet's tail.

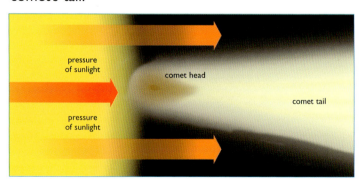

pressure of sunlight

comet head

comet tail

pressure of sunlight

The tail always points away from the Sun, so when the comet orbits toward the Sun, its tail follows behind. But when the comet moves away from the Sun in its orbit, its tail streams out in front of it.

INTO THE DEEP FREEZE

As the comet gets farther away from the Sun, it grows colder. Not as much ice melts, and not as much gas and dust are released. The head and tail start to shrink and gradually fade. After a while, the comet freezes completely. There is no cloud to reflect sunlight. Once again, the comet becomes invisible to us. It will not be seen again until it next closes in on the Sun—in some cases, not for thousands of years.

Comet West, the brightest comet of 1976, was easily visible to the naked eye. Astronomer Richard West had discovered the comet the previous year.

WHERE COMETS CAME FROM

Comets are ancient remains of ice and dust left over after the planets formed. Astronomers think that most comets collected in a "cloud" at the outer reaches of the solar system, where billions of them remain. This comet storehouse is called the Oort Cloud, and it surrounds the solar system. Many of the comets that appear in our skies come from the Oort Cloud, but others come from an area closer to the Sun, called the Kuiper Belt. This belt is just beyond Neptune's orbit, near Pluto.

STARPOINT

Every time a comet travels around the Sun, it can lose as much as 500 million tons of material.

FAMOUS COMET

Halley's comet, also called Comet Halley, is the most famous comet. People have spotted it every time it has returned to Earth's skies since 240 B.C. It was the comet that appeared at the time of the Battle of Hastings in A.D. 1066. It last appeared in 1986, but it was not easy to see because it was so faint.

The comet is named after the English astronomer Edmond Halley. He saw it in 1682 and, after checking past records, realized it was similar to comets seen in 1531 and 1607. He guessed that it was the same comet, and he predicted that the comet would return again 76 years later, in 1758. It did, 16 years after Halley had died.

At its last appearance, in 1986, Halley's comet was a faint object in the night sky.

This ancient woodcut shows the bright comet of 1456, which we now know was an appearance of Halley's comet.

Meteoroids, Meteors, and Meteorites

Meteors appear as streaks of light in the night sky. Occasionally, they fall through our skies and land on Earth's surface as meteorites.

Outer space is full of specks of rock and metal. These small pieces of material are called meteoroids. Like asteroids and comets, meteoroids orbit the Sun.

On its own journey through space, Earth comes across these orbiting pieces all the time. It attracts nearby meteoroids with the pull of its gravity. These tiny specks of matter leave bright trails behind them as they travel through Earth's atmosphere, or the layer of gases around the planet. From Earth, these bright trails look like stars falling from the sky. On a clear night, you might be able to see several "falling stars" every hour. Scientists call these trails of light meteors.

A meteor burns a trail through Earth's atmosphere. The flash near the end of the trail shows where the meteor has exploded.

This woodcut from 1557 illustrates a meteor shower as a battle in the heavens.

BURNING UP

As a meteoroid falls toward Earth, it enters Earth's atmosphere. The friction, or rubbing, of Earth's atmosphere against the surface of the falling particle causes it to heat up. As it gets hotter and hotter, it starts to burn. When a meteoroid leaves outer space and begins to burn through Earth's atmosphere, it is called a meteor. As it burns, the meteor leaves a flaming trail behind it, until it burns away to nothing. A meteor's bright trail is often easy to see in our night sky.

The burning trails of most meteors last only a few seconds. Many meteors are no bigger than a grain of sand, and their small size causes them to burn up quickly. Occasionally, a larger meteor enters Earth's atmosphere. It burns brighter and for a longer period of time. An especially bright meteor is called a fireball.

Some meteors burn up completely as they plunge through the atmosphere. A few are big enough to survive and reach the ground.

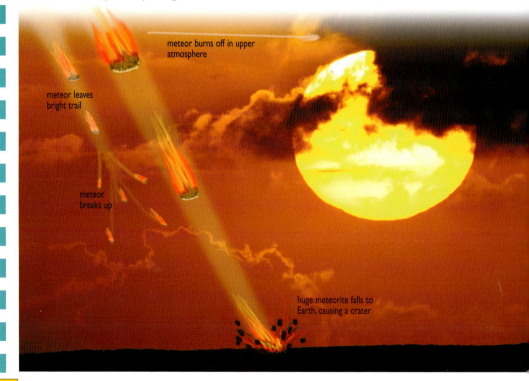

meteor burns off in upper atmosphere

meteor leaves bright trail

meteor breaks up

huge meteorite falls to Earth, causing a crater

SHOWERS OF METEORS

At certain times of the year, many meteors appear to rain down at once in the night sky. Scientists call this increased meteor activity a meteor shower. Meteor showers occur when Earth passes through tiny pieces of rock and metal left behind by an orbiting comet. These clusters of leftover pieces are drawn into Earth's atmosphere, where they burn up and streak through our skies.

Regular showers take place every year in a certain part of the sky. They are named after the constellation, or pattern of stars, that the meteors appear to come from. The Perseids, which occur in August, appear to come from the constellation Perseus. The Orionids, in October, seem to come from the constellation Orion. Earth passes through these clusters of comet particles around the same time each year, so astronomers can predict when the meteor showers will occur.

STARPOINT

Meteor particles hit Earth's atmosphere at speeds of up to 150,000 miles (250,000 km) an hour.

In the nineteenth century, scientists sometimes used hot air balloons to get a better view of events in the night sky. This Leonid shower was observed in 1870.

Meteor Storm

The Leonids, a meteor shower that appears to come from the constellation Leo, occurs each year in November. Chinese astronomers saw the first Leonid meteor shower in A.D. 902. They thought the stars were falling like rain. The 1966 Leonid shower was one of the greatest meteor showers on record. So many meteors fell, it was called a meteor storm. At one point during the storm, some observers saw 40 meteors each second!

Above: This is the world's biggest meteorite, found in southern Africa. It fell to Earth in prehistoric times.

METEORITES

Sometimes Earth crosses paths with a much larger meteoroid than usual. The meteoroid drops through the atmosphere and starts burning as a meteor. But if it is big enough, only its outer layer will burn away. The rest of the meteor will fall to the ground. When a meteor has fallen through Earth's atmosphere and lands on Earth, it is called a meteorite. Meteorites weighing thousands of pounds have been found all over the world. The biggest, the Hoba meteorite, was found in Namibia in southern Africa. It weighs about 120,000 pounds (54,000 kg).

There are three main kinds of meteorites. The largest group, stony meteorites, are made up mainly of rocky materials. They account for over 90 percent of all meteorites that land on Earth. A second kind are made of iron and are called iron meteorites. About 1 percent of meteorites contain nearly equal amounts of stone and metal. They are called stony-iron meteorites.

These two meteorites were found in Antarctica. Astronomers think that one (above) was probably broken off from Mars and the other (right) was broken off from the Moon.

WHERE METEORS CAME FROM

Where did these specks and lumps of rock and metal come from? Some meteors and meteorites are probably pieces of the Moon and other planets, such as Mars. Smaller pieces from these larger bodies were broken off in collisions with asteroids. Other meteors and meteorites probably came from the asteroid belt. They formed when asteroids slammed together and broke apart into smaller pieces. Still others are probably dust particles left behind by comets.

STARPOINT

The largest meteorite in a museum is the Ahnighito, located in the American Museum of Natural History in New York. Explorer Robert Peary found it in Greenland in 1897.

DIGGING CRATERS

Most meteorites are slowed down as they travel through Earth's atmosphere. By the time they reach Earth's surface, their speed and size are not great enough to hit Earth with much force. Only the largest meteorites dig out craters, or large pits, when they hit Earth's surface.

Scientists believe there are over 100 meteor craters on Earth. In the past there were many more. But over millions of years, ancient craters have been worn away by the process of erosion. Erosion is the gradual wearing away of Earth's surface by natural elements such as wind or flowing rivers.

This map shows the locations of known craters made by meteorites.

Barringer Crater in the Arizona desert is the most famous and best-preserved meteorite crater on Earth.

Mercury's surface is covered with craters. It was battered by meteorites billions of years ago.

THE WORLD'S BIGGEST CRATER

The best-preserved and largest meteorite crater on Earth is the Barringer Crater, also called Meteor Crater, in Arizona. It is a great pit in the Canyon Diablo region of the Arizona desert, located about 20 miles (32 km) west of the town of Winslow. It measures more than 4,000 feet (1,200 m) across and about 600 feet (180 m) deep.

The crater is named after Daniel Barringer, a silver miner in the early 1900s. He was one of the first people to be convinced that a meteorite had formed the

crater. He drilled into the crater floor and found a layer containing pieces of nickel. The nickel pieces were remains of the meteorite that had blasted out the crater.

Scientists believe that this meteorite fell to Earth about 50,000 years ago. It would have been the size of a railroad car and weighed as much as 300,000 tons. No large pieces of the meteorite have ever been found. But this is not surprising. The meteorite would have hit the ground so hard that it would have smashed to pieces. All that remains of the monster meteorite are many pieces of iron found in the surrounding region.

SHAPING PLANETS AND MOONS

Many other planets and moons in the solar system have also been shaped by collisions with meteorites. Most collisions took place billions of years ago, not long after the planets and their moons formed.

You can see the result of these ancient collisions if you look at the Moon. Much of its surface is covered with craters large and small. The craters have barely changed since they were formed. Because the Moon does not have any atmosphere to create weather, there has been no erosion to wear away the craters. Among the planets, Mercury is the most heavily cratered body. Like the Moon, it has never had an atmosphere, so there has been no erosion to change its surface. Venus and Mars have fewer craters because, like Earth, both planets have an atmosphere.

Meteorites created most of the craters on the Moon (above). A meteorite impact made the large crater on Mars named Argyre Planitia (below).

Visiting Comets and Asteroids

For years, astronomers had only been able to study comets and meteors from far away. Space probes have given us our first glimpses of these small worlds from close up.

In 1986, Halley's comet returned to Earth's skies for the first time since 1910. As the comet passed closest to Earth in March 1986, five space probes were sent to study it. They were launched by Russia, Japan, and Europe. The probes were part of a worldwide investigation known as International Halley Watch, in which over 800 scientists in 40 countries took part.

Two Russian probes, Vega 1 and Vega 2, were launched in December 1984. In January 1985, the first of two Japanese probes was launched. It was named Sakigake, meaning pioneer. This was a good name because Sakigake was the first probe Japanese space scientists had ever launched. In July, the European Space Agency (ESA) launched its probe, Giotto. Japan's second probe, Suisei, meaning comet, followed in August.

THE ENCOUNTERS

The two Russian probes, Vega 1 and Vega 2, were the first to encounter, or meet, Halley's comet. They took pictures, measured the amount of dust around the comet, and investigated the contents of the comet's nucleus. Scientists learned

Above: Vega 1 and Vega 2, similar to the spacecraft above, were the first probes to study Halley's comet.

Right: The Suisei spacecraft was small compared to the Vega. It weighed only about 300 pounds (140 kg), compared to Vega's 5 tons.

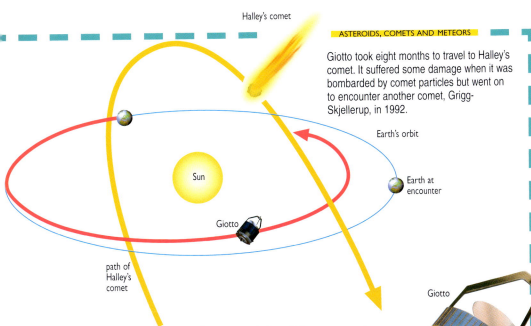

Halley's comet

Giotto took eight months to travel to Halley's comet. It suffered some damage when it was bombarded by comet particles but went on to encounter another comet, Grigg-Skjellerup, in 1992.

Earth's orbit

Sun

Earth at encounter

Giotto

path of Halley's comet

Giotto

that Halley's comet is made up mainly of frozen water and frozen carbon dioxide, or dry ice.

The two Japanese probes, Suisei and Sakigake, made useful observations on the effect of the comet on the solar wind, the stream of particles coming from the Sun. And Suisei took pictures of bright gas jets coming from the comet.

Europe's Giotto was the last probe to encounter Halley. It flew past the comet at a distance of only about 370 miles (600 km), the closest of all the probes. It took pictures that showed the bright gas jets and also the comet's nucleus. The nucleus was dark in colour and only about 10 miles (16 km) across.

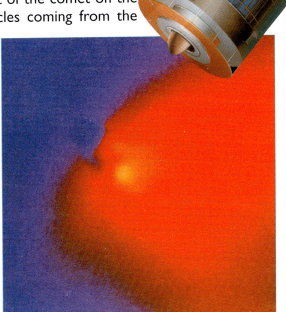

Giotto snapped this picture of gas jets shooting out from the nucleus of Halley's comet.

Above: Galileo's on-board rocket engine fires after the probe has been placed in orbit by space shuttle Atlantis on October 18, 1989.

VISITING ASTEROIDS

In October 1989, Galileo was successfully launched by NASA. Galileo left Earth on a roundabout route that would eventually take it to Jupiter. NASA scientists planned for it to pass twice through the asteroid belt and take close-up pictures of asteroids for the very first time.

Galileo followed a long looping path through the solar system. It first flew near Venus, and it was speeded up by the pull of Venus's gravity. Then it flew back to Earth, where it was again speeded up by Earth's gravity. It then sped out toward the asteroid belt. In October 1991, Galileo turned its instruments on an asteroid named Gaspra, which proved to be irregularly shaped and heavily cratered.

Galileo next looped back toward Earth to gather even more speed before returning to the asteroid belt. In August 1993, its target was a larger asteroid called Ida. Scientists were amazed to discover that Ida had a tiny moon circling around it. They named the moon Dactyl.

Next Galileo continued through the asteroid belt to its main target, Jupiter, where it would explore the giant planet's atmosphere and moons.

Left: This lump of rock is the asteroid Ida. Galileo photographed the asteroid in 1993 on its way to Jupiter. Ida measures about 35 miles (55 km) long.

Bombarding Earth

Huge chunks of rock from outer space have bombarded Earth in the past and may do so again in the future.

If an object the size of the meteorite that created Barringer Crater hit New York City, the city would be flattened. Much larger meteorites and asteroids have hit Earth in the more distant past. They have not only torn out great craters, they have also affected weather conditions and wiped out many plant and animal species.

A collision with a huge asteroid may have been the reason dinosaurs died out about 65 million years ago. Some scientists suggest that when this large asteroid hit Earth, it blasted millions of tons of rock and dust into the air. These particles quickly spread through the atmosphere, covering Earth with a dust cloud. This layer of dust was so thick and black that light from the Sun could not get through. Without sunlight, plants would have stopped growing, because plants need sunlight to make their food. Since most animals rely on plants for food, many of them, including the dinosaurs, would have died out.

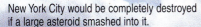

New York City would be completely destroyed if a large asteroid smashed into it.

Comet Collisions

Not all collisions with objects from outer space have caused destruction on Earth. Some might have helped bring about life. Billions of years ago, Earth did not have all of the elements needed to produce and maintain life. It is possible that icy comets crashing into Earth brought water and other raw materials to our planet. Over time, these materials may have helped make it possible for life to exist on Earth.

Glossary

asteroid: a rocky body that orbits the Sun between Mars and Jupiter

asteroid belt: a ring-shaped region in the solar system, between Mars and Jupiter, in which most asteroids are found

atmosphere: the layer of gases around a planet or moon

atmospheric pressure: the force of the gases in an atmosphere pressing down

aurora: a glow produced in the polar regions of some planets by solar wind particles entering the atmosphere

axis: an imaginary line through a planet from its north to its south pole

belt: a dark-coloured band in Jupiter's atmosphere

captured rotation: when a moon rotates on its axis in the same amount of time that it takes to orbit its planet once

chasma: a crack, or valley, on the surface of a planet or moon

co-orbital: a moon that travels in the same orbit as another moon

comet: a small body, made up of dust and ice, that orbits the Sun and shines when it nears the Sun

core: the centre part of a planet or moon

crater: a pit on the surface of a planet or moon

crust: the hard surface of a rocky or icy planet or moon

double planet: a planet and moon system in which the moon is relatively large, as with Pluto and its moon Charon

equator: an imaginary line around the centre of a heavenly body

Galilean moons: the four large moons of Jupiter, discovered by Galileo

gas giant: a large planet that is made up mainly of gas; Jupiter, Saturn, Uranus, and Neptune are the gas giants in our solar system

gravity: the attraction, or pull, that a heavenly body has on objects on or near it

Great Red Spot: a huge rotating storm system on Jupiter

interplanetary space: the space between the planets

jet stream: a fast-moving air current in a planet's atmosphere

magnetosphere: the region in space around a planet where its magnetism can be detected

mass: the amount of matter in a body

meteorite: a lump of rock or metal from space that hits a planet or a moon

moon: a natural satellite of a planet

morning star: the planet Mercury or Venus shining in the eastern sky just before sunrise

NASA: the National Aeronautics and Space Administration, which organizes space activities in the United States

orbit: the path in space of one heavenly body around another, such as Jupiter around the Sun

probe: an unmanned spacecraft that travels from Earth to one or more heavenly bodies

ring system: a set of rings found around a giant planet, made up of fine particles or lumps of rock and ice

ringlet: a very narrow ring

scarp: a steep cliff

shepherd moon: a tiny moon located near a planet's ring that may help keep the ring particles in place

solar system: the Sun and all the bodies that circle around it, including Saturn and the other planets

solar wind: a stream of particles given off by the Sun

transit: the time when Mercury or Venus travels across the face of the Sun, as viewed from Earth

Trojans: groups of asteroids that travel in the same orbit as Jupiter

water vapour: water in the form of gas

zone: a light-coloured band in Jupiter's atmosphere

Index